AFRICAN POLITICAL, ECONOMIC, AND SECURITY ISSUES

NORTH AFRICA

SOCIAL, ENVIRONMENTAL AND POLITICAL ISSUES

African Political, Economic, and Security Issues

Additional books in this series can be found on Nova's website under the Series tab.

Additional e-books in this series can be found on Nova's website under the eBooks tab.

AFRICAN POLITICAL, ECONOMIC, AND SECURITY ISSUES

NORTH AFRICA

SOCIAL, ENVIRONMENTAL AND POLITICAL ISSUES

STEPHEN FULLER
EDITOR

nova
science publishers
New York

Library of Congress Cataloging-in-Publication Data

ISBN: 978-1-53612-983-0

Published by Nova Science Publishers, Inc. † New York

CONTENTS

PREFACE

In *North Africa: Social, Environmental and Political Issues*, the authors begin by presenting that the precursors of an intense coupled system of an African easterly wave and a mesoscale convective system over the lee of Darfur Mountains around 1800 UTC, August 12, 2004 were traced back to the southern Arabian Peninsula, Asir Mountains, and Ethiopian Highlands using gridded satellite data, European Centre Medium-Range Weather Forecast Intermediate data, and the Advanced Research Weather Research and Forecasting model simulated data. Anticipated patterns of future energy use and consequent environmental impacts are discussed, along with potential solutions to current problems with the adoption of the Sahara Solar Breeder project plan. The relations between renewable energy and sustainable development are described with practical cases of the Sahara Solar Breeder energy policy and its portfolio in the context of continental and global clean energy supply. The authors seek to answer a question: "Will the new technologies that have sustained globalization reinforce or undermine democracy?"

Chapter 1 - The precursors of an intense coupled system (A04) of an African easterly wave (AEW) and a mesoscale convective system (MCS) over the lee of Darfur Mountains around 1800 UTC, August 12, 2004 were traced back to the southern Arabian Peninsula, Asir Mountains, and Ethiopian Highlands using gridded satellite (GridSat) data, European Centre Medium-Range Weather Forecast Intermediate (ERA-I) data, and the Advanced Research Weather Research and Forecasting (WRF) model simulated data. A vorticity budget analysis was performed to investigate the dynamics and mechanisms that contributed to the enhancement of A04's vorticity perturbation. It was found that convergence of southeasterly and northwesterly wind from anticyclones over Arabian and Indian Ocean generates cyclonic vorticity over the southern Arabian Peninsula that was advected by an easterly wind maximum embedded within the African easterly jet. Convergence and vorticity advection were the most important factors to the formation of A04's vorticity perturbation, while cloud clusters were generated by diurnal heating over the Asir Mountains, Ethiopian Highlands and Darfur Mountains. Both vorticity perturbation and cloud clusters went through their formation and propagation phases, intensified by lee vortices, and then eventually merged into an AEW-MCS system on the lee of the Darfur Mountains, consistent with the conceptual model proposed in Lin et al. (2013).

Chapter 2 - Global energy consumption is projected to increase, even in the face of substantial declines in fossil energy intensity, at least 2-fold by mid-century relative to the present because of population and economic growth. Currently the world's growing thirst for oil amounts to almost 1000 barrels a second. The ever increasing world energy demand cannot be satisfied much longer with fossil fuels. Achieving solutions to energy problem that we face today requires long-term potential actions for sustainable development. In this regard, renewable energy (RE) resources appear to be crucial and the one of the most efficient and effective solutions for sustainable development. That is why a shift to a more sustainable mix by means of the use of RE sources, an increase of the energy efficiency, energy sobriety and greenhouse gas mitigation measures is needed. The development of technological innovation in the energy fields is central to the success of the energy paradigm shift worldwide. Among RE resources, solar energy is by far the largest untapped resource and major primary energy exploitable resource. Algeria has the highest average solar radiation per square meter of any country in the world and is an interesting case study in that it had quite a slow start in this field but is now forging ahead with RE –at least at the utilities level, boosted by government policy. Progress at residential level is not as fast, though. This has to be rectified for the country to meet the targets set in its National Development Plan. A broad range of intensive and promising research and development is being conducted, under the Sahara Solar Breeder (SSB) project concept, in order to produce technological options that can allow energy security, climate stabilisation and economic development from solar energy. In view of the intermittency of insolation, SSB project provides an especially attractive approach in which solar energy must be dispatched on demand to the end user using High Critical Temperature Supra Conductor (HT_cSC) cables. The magnitude of such implied infrastructure transition suggests the need for massive investments in innovative energy research. Anticipated patterns of future energy use and consequent environmental impacts are comprehensively discussed in this article. Also, potential solutions to current problems are identified along with the adoption of the SSB project plan and activities. The relations between RE and sustainable development are described with practical cases of SSB energy policy and its portfolio directed towards continental and global clean energy supply with existing solar grade Silicon production from Sahara sand and diatoms technologies making photovoltaic (PV) a significant contributor within a portfolio of energy sources in the coming 10 to 50 years. An illustrative example of some developed SSB key innovative technologies is presented and more should be challenged together, Algerian and Japanese SSB research teams, to seek the paradigm shift in global energy production and transport system. In both cases, the research is emanating from the University of Sciences and Technology of Oran (USTO-MB) in line with Saida University.

Chapter 3 - North Africa is marked with an unmatched phenomena of increase in IT, though IT users are not homogenous groups, but rather heterogeneous communities with diverse interests and motivations. As explained by the Swiss Agency for Development and Cooperation (SDC) that technology itself cannot solely make a difference in the practice of the sociopolitical and economic progress, but rather the other way around where indicators of development behind the technology drives the change. IT is thus an instrument, not a goal.

The synergy between politics and IT is exemplified through many hybrid formations and phases since its initial preparation phase, where activists used digital media over several years to build social networks and identified collective identities. The second launching phase is marked with unprecedented ignition moments, where movement leaders activated these digital networks toward offline protests. And thirdly, profound channeling and programming of information warfare are in place to compete for control, as well as enhance international pressure, economic sanctions, and lobbying efforts.

But there are a number of concerns and limitations to what we know, expect and do with IT and digital democratization which coincides with the words of Schraeder (2003), when he described Africa's diversities and complexities as "mosaic in transformation." Such complex and inconclusive crossing roads are further interoperated to serve commercial and political interests in what is called the "Rashmon Effects".

This makes it rational to mention that the level of legitimacy in using power and establishing identity is the main prevailing trend in Africa, which exploits the social function of media as a driving force for national ideology. Social power in many African societies construct meanings in the human mind through communication that utilizes internet and other horizontal digital communication networks.

This research puts forward a key question: "Will the new technologies that have sustained globalization reinforce or undermine democracy?" The aim is to offer a more nuanced approach of deductive forms of investigation about the potential role and impact of digital politics. This difficult path though must be considered by cross examining how politics, digital infrastructure adoption function and affect could contribute to democratization within the African context.

In: North Africa
Editor: Stephen Fuller
ISBN: 978-1-53612-983-0

Chapter 1

FORMATION AND MAINTENANCE OF AN AFRICAN EASTERLY WAVE MESOSCALE CONVECTIVE SYSTEM IN 2004 OVER EAST AFRICA AND THE ARABIAN PENINSULA

Yuh-Lang Lin[1,*], James Spinks[2] and Bo-Wen Shen[3]
[1]Department of Energy and Environmental Systems and Department of Physics, North Carolina A&T State University Greensboro, NC, US
[2]College of Technology, ECPI University, Greensboro, NC, US
[3]Department of Mathematics, San Diego State University, San Diego, CA

ABSTRACT

The precursors of an intense coupled system (A04) of an African easterly wave (AEW) and a mesoscale convective system (MCS) over the lee of Darfur Mountains around 1800 UTC, August 12, 2004 were traced back to the southern Arabian Peninsula, Asir Mountains, and Ethiopian Highlands using gridded satellite (GridSat) data, European Centre

* Corresponding Author Email: ylin@ncat.edu.

Medium-Range Weather Forecast Intermediate (ERA-I) data, and the Advanced Research Weather Research and Forecasting (WRF) model simulated data. A vorticity budget analysis was performed to investigate the dynamics and mechanisms that contributed to the enhancement of A04's vorticity perturbation. It was found that convergence of southeasterly and northwesterly wind from anticyclones over Arabian and Indian Ocean generates cyclonic vorticity over the southern Arabian Peninsula that was advected by an easterly wind maximum embedded within the African easterly jet. Convergence and vorticity advection were the most important factors to the formation of A04's vorticity perturbation, while cloud clusters were generated by diurnal heating over the Asir Mountains, Ethiopian Highlands and Darfur Mountains. Both vorticity perturbation and cloud clusters went through their formation and propagation phases, intensified by lee vortices, and then eventually merged into an AEW-MCS system on the lee of the Darfur Mountains, consistent with the conceptual model proposed in Lin et al. (2013).

1. INTRODUCTION

Earlier studies showed that African easterly waves form between May and October when large-scale environmental conditions favor the existence of the African jet. African easterly waves may propagate westward across the tropical and subtropical North Atlantic and reach the Caribbean Sea and western North Atlantic. Some African easterly waves become hurricanes. In particular, observational analyses indicated that AEWs possess the following basic characteristics (Reed et al., 1977; Burpee, 1974; see review in Chen, 2006): (1) a propagation speed of 7 – 9 ms^{-1}, (2) a wavelength of ~2000 – 4000 km, and (3) propagating along the rainy zone and to the south of the African easterly jet (AEJ) around 10°N (AEW_s) or along the Saharan thermal low near 20°N (AEW_n).

Several mechanisms have been proposed for the formation of AEWs in previous studies: (1) release of barotropic-baroclinic instability (Charney-Stern 1962, Burpee 1972, Chen 2006), which is more applicable to the AEW_s; (2) release of baroclinic instability (Chang 1993; Thorncroft 1995) which is more applicable to the AEW_n. Also included is the release of the proclaimed African Easterly Jet (AEJ) instability (e.g., Simmons 1977; Thorncroft and Hoskins 1994a, b; Grist et al., 2002), which is assisted by the baroclinicity associated with the Sahara Desert; (3) latent heating (e.g., Hsieh & Cook 2005, 2007; Thorncroft et al. 2008); and (4) orographic forcing (Mozer & Zehnder 1996; Lin et al., 2005, 2013; Mekonnen et al. 2006; Laing et al. 2012). In later work of pre-Debby (2006) AEW-MCS (Lin et al., 2013), a conceptual model (Figure 1) was proposed to illustrate the role of the interaction between northeasterly Shamal wind and southwesterly wind from the India Ocean anticyclone in producing cyclonic vorticity perturbation, which served as the precursor of the AEW, over the southern Arabian Peninsula. More details on the conceptual model are provided below.

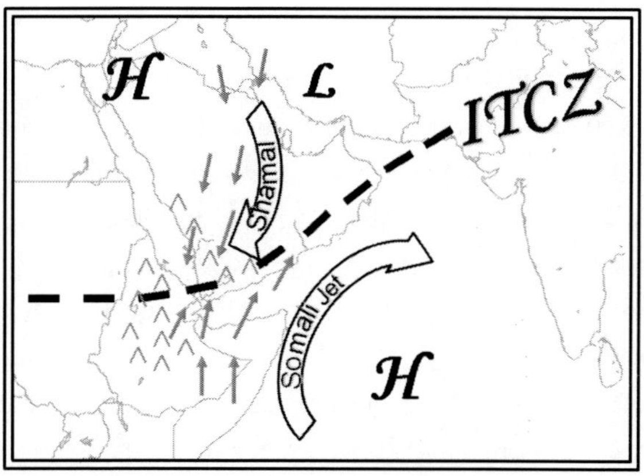

Figure 1. A conceptual model of the generation of cyclonic vorticity perturbations and convective cloud clusters preceding the pre-Debby (2006) AEW–MCS system, as proposed in Lin et al. (2013). The sources of the convective cloud clusters and vorticity perturbations were attributed to the cyclonic convergence of northeasterly Shamal wind and the Somali jet, especially when the Mediterranean High shifted toward east with high pressure ridge extended farther to the southeast and the Indian Ocean high strengthened and its associated Somali jet penetrated farther to the north. The cyclonic vorticity perturbations were strengthened by the vorticity stretching associated with convective cloud clusters and the convective cloud clusters are strengthened by the diurnal convection over the Sarawat Mountains, Asir Mountains, Ethiopian Highlands, and Darfur Mountains downstream of the genesis region (i.e., southwest Arabian Peninsula).

AEWs are often associated with mesoscale convective systems (MCS) while they travel westward (e.g., Payne & McGarry, 1977; Laing & Fritsch, 1993; Fink & Reiner, 2003). This helps alter convection and rainfall over the African continent daily (e.g., Carlson, 1969) and makes AEWs detectable from satellite imagery. MCSs tend to occur in preferred regions of AEWs: at or ahead of the AEW trough in the vicinity of the intertropical convergence zone (ITCZ), and collocated with the trough over the ocean. Hill and Lin (2003) and Lin et al. 2005, back-tracked the pre-Alberto AEW-MCS (2000) precursor to the Ethiopian Highlands, while Mekonnen et al. (2006) found that the Darfur Mountains are particularly important for providing convective precursors that propagate westward and trigger AEWs downstream. Although the MCS has been used as proxy for detecting the AEWs, there are distinct features of the convective system and the wave perturbation (e.g., Lin et al. 2013). Thus, it is important to trace both of them or their individual precursors near their source regions.

Based on statistical analysis of AEW activities, Hodges and Thorncroft (1997) showed that northern Africa could be divided into areas of genesis or lysis regions for growing or dissipating convection, respectively. The areas are determined by using the coldest cloud-top temperatures to track several MCSs across northern Africa where the appropriate convective activities were identified. Other studies have also shown that most MCSs formed in baroclinic zones and downstream of mountain ranges (Laing & Fritsch, 1993; Laing et al., 2012), such as the case over northern Africa during the Northern Hemisphere summer months.

The pre-tropical storm Debby (2006) AEW (also denoted as pre-Debby AEW hereafter) was traced back to a vorticity genesis region in eastern North Africa by Lin et al. (2013). This region is located over the southern Arabian Peninsula and was investigated to see formation and evolution of the pre-Debby AEW-MCS. In this study, we will concentrate on the origin of a coupled *AEW-MCS in August of 2004* (denoted as A04, hereafter) while taking a similar approach that was used to explain vorticity generation in Lin et al. (2013). A04's MCS/convective clusters and A04's AEW/vorticity perturbation will be examined to understand the mechanisms that generate these systems. The reason for investigating this

A04 AEW-MCS is that its characteristics are very similar to those of pre-Debby AEW-MCS, but did not trigger tropical cyclone over the Atlantic.

Eastern North Africa and the Arabian Peninsula include mountains that will have an effect of the generated vorticity and MCSs, especially during diurnal cycles. The influence of orography has been observed in the development of lee vortices and, subsequently to AEWs (Hill and Lin 2003). This study will focus on the analysis of wind and vorticity fields along with diabatic processes to identify A04's origins to verify the conceptual model proposed in Lin et al. (2013; Figure 1) for the generation of AEW precursor (cyclonic vorticity perturbation) and convective cloud clusters preceding the pre-Debby (2006) AEW–MCS system. In Lin et al.'s conceptual model, the sources of the convective cloud clusters and vorticity perturbations are attributed to the cyclonic convergence of northeasterly Shamal wind and the Somali jet, especially when the Mediterranean High shifts toward east with high pressure ridge extended farther to the southeast and the Indian Ocean high strengthened and its associated Somali jet penetrated farther to the north. The cyclonic vorticity perturbations are strengthened by the vorticity stretching associated with convective cloud clusters and the convective cloud clusters are strengthened by the diurnal convection over the Sarawat Mountains, Asir Mountains, Ethiopian Highlands, and Darfur Mountains downstream of the genesis region, i.e., the southwest Arabian Peninsula.

The rest of this paper is organized as follows: section 2 describes data used and the experiment design for the numerical case study. Section 3 presents the results with two subsections, including observational analysis from satellite and reanalysis data in section 3.1 and numerical results from the model in section 3.2. Concluding remarks are provided at the end of the paper.

2. Data and Experiment Design

In order to investigate the origins and propagation of the AEW and MCS, various datasets were used to identify case study A04. For AEW convection analysis, observational globally gridded satellite (GridSat) data from the National Climatic Data Center (NCDC) at an 8-km spatial resolution and 180-min temporal resolution (Knapp et al., 2011) were used. The GridSat data that serve as reliable observations for cloud clusters are analyzed along with the global European Centre Medium-Range Weather Forecast Intermediate (ERA-I) dataset at a resolution of 0.75° x 0.75° and a time interval of 6 h (Dee et al. 2011). Using the ERA-I, analysis was performed to track AEWs and MCSs and investigate the regions of genesis. Daily precipitation associated with A04 is illustrated using the data from the National Oceanic and Atmospheric Administration Climate Prediction Center's Famine Early Warning System (NOAA CPC FEWS). Since MCSs are associated with precipitation, the NOAA CPC FEWS data may provide information on MCSs and can be inter-compared with the GridSat and ERA-I to track A04's MCS.

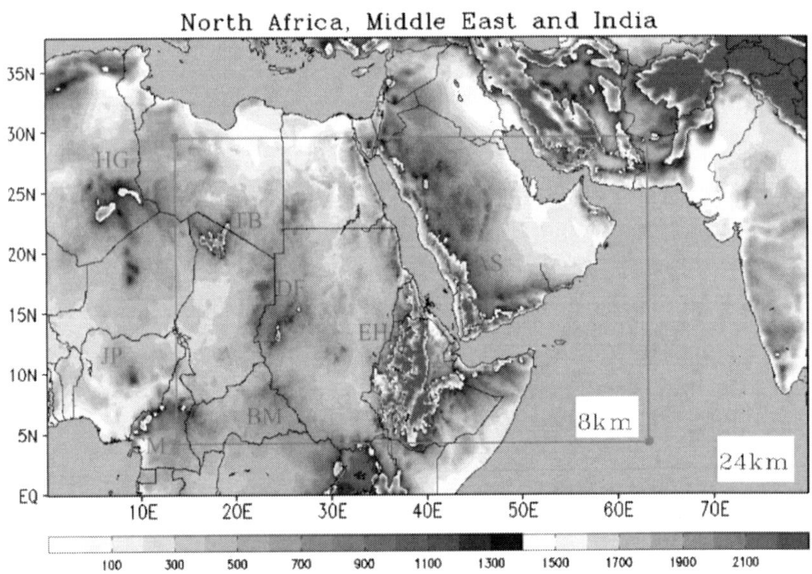

Figure 2. WRF simulated geography for Africa (north of 0°) and southern Arabia Peninsula. Labels indicate major geographic features: *AS* Asir Mountains, *BM* Bongo Massif, *CM* Cameron Highlands, *DF* Darfur Mountains, *EH* Ethiopian Highlands, *HG* Hoggar Mountains, *JP* Jos Plateau, and *TB* Tibesti Mountains. The outer domain of the simulation is 24 km and the inner domain (red lines) is 8 km resolution.

Once a case study is verified from observational analysis, the Advanced Research Weather Research and Forecasting (WRF) model version 3.3.1 (Skamarock et al. 2008) will be used to simulate the phenomena. The WRF model is a numerical weather prediction (NWP) and atmospheric simulation system designed for both research and operational applications. The WRF model is a compressible, three-dimensional, nonhydrostatic model using terrain-following coordinates. The governing equations for WRF are written in flux-form with conserved mass and dry entropy. In the numerical model, the third-order Runge-Kutta time scheme was employed and fifth and third order advection schemes were chosen for the horizontal and vertical directions, respectively. An open (radiative) lateral boundary condition was used in the north-south direction, a free-slip condition was used for the lower boundary condition, and a periodic boundary condition was used in the east-west direction.

Figure 2 illustrates the domains used in WRF to simulate A04. The 24-km outer domain consist of 350 x 180 grid points, 38 vertical levels, and 120 second time steps. The 8-km inner nested domain consist of 652 x 352 grid points, 38 vertical levels, and 40 second time steps. The physics options for both domains include: Goddard microphysics, RRTM longwave radiation, Duhdia shortwave radiation, Mellor-Yamada-Janjic PBL physics, and Betts-Miller-Janjic cumulus parameterization.

3. Results

3.1. Analysis from Observations and ERA-I

3.1.1. Tracking the Convection of A04

A04 was identified by analyzing Hovmoller diagrams from Gridsat brightness temperature and comparing it with ERA-I cloud cover and relative vorticity. The idea of this comparison was to note regions of intense easterly waves that originated from the Arabian Peninsula (50°E) and propagated into East Africa. Figure 3 illustrates A04's MCS that was formed at 0600 UTC, August 9, 2004 (for simplicity, it is denoted as

09/06Z hereafter). The Gridsat and ERA-I's cloud cover compares well, thus giving confidence in the simulated MCS from the ERA-I is realistic enough to proceed with the case study of A04.

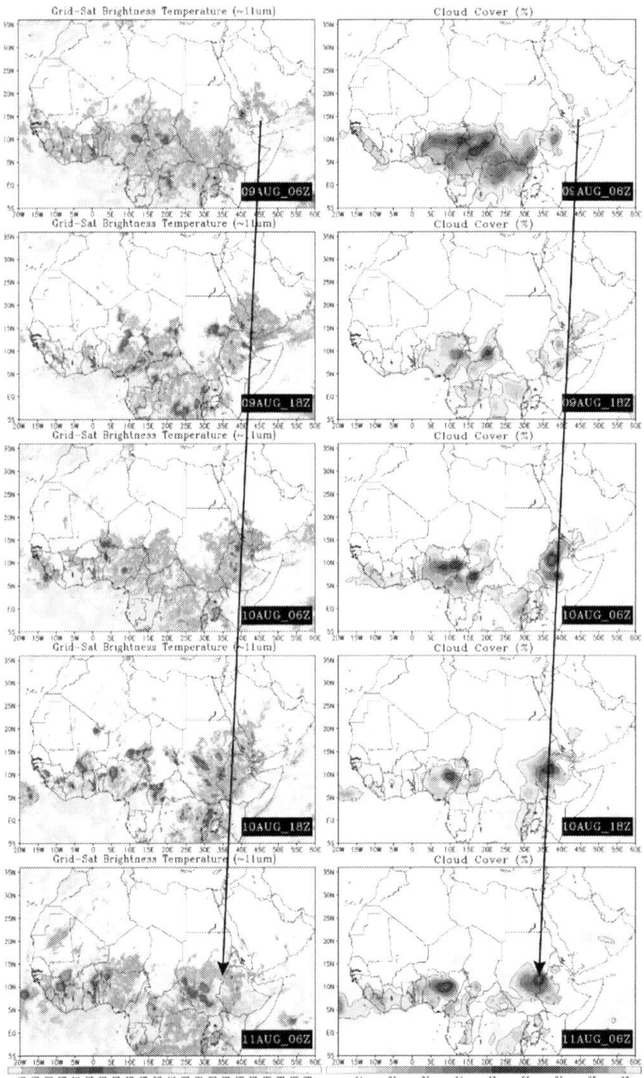

Figure 3. (a) Gridsat (left column) brightness temperature versus ERA-I (right column) total cloud cover (%). Black line denotes track of case study A04 between 08/09/06Z to 08/11/06Z every 12h.

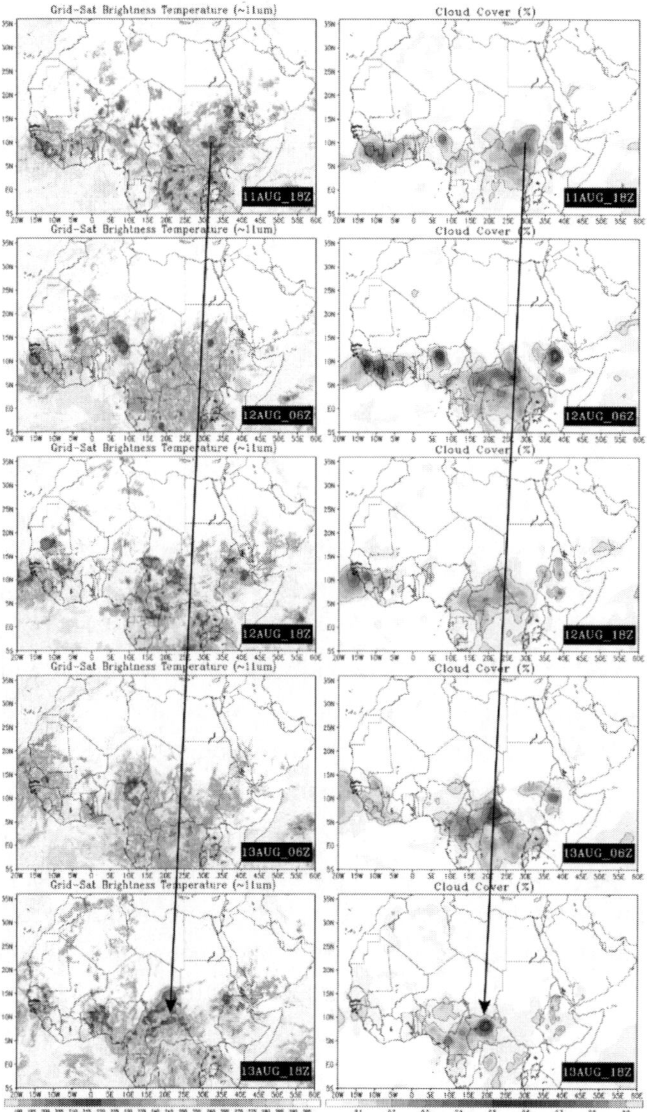

Figure 3. (b) Gridsat (left column) brightness temperature versus ERA-I (right column) total cloud cover (%). Black line denotes track of case study A04 between 08/11/18Z to 08/13/18Z every 12h.

In Figure 3a, the associated MCS for A04 can be seen in both from Gridsat and ERA-I at 9/06Z over the southern Arabian Peninsula. The

MCS propagates over the Asir Mountains at 9/18Z and the Ethiopian Highlands at 10/06Z where orographic effects will enhance the convection. At 10/06Z, there is a noticeable convective cloud burst over the Ethiopian Highlands. The MCS begins to weaken in the Gridsat at 11/06Z, but remains intense in the ERA-I. In Figure 3b at 11/18Z, the MCS propagates away from the Ethiopian Highlands crossing Sudan at 12/06Z and 12/18Z. In the Gridsat at 13/06Z the MCS has entered a dissipation phase, before there is another convective cloud burst of the Darfur Mountains and 13/18Z. The ERA-I does not show dissipation of convection at 13/06Z, but did resolve the convective cloud burst at 13/18Z.

Overall, the ERA-I compared well with convection from Gridsat, which showed the propagation of A04's associated MCS. Maintenance of convection based on satellite can be attributed to orographic effects from the Asir Mountains, Ethiopian Highlands, and Darfur Mountains, especially during diurnal cycles.

To further strengthen the confidence of A04 from the ERA-I being realistic, the NOAA CPC FEWS daily precipitation is plotted with ERA-I 600 mb geopotential heights (Figure 4). On the 8th and 9th of August over the southern Arabian Peninsula, an inverted trough can be seen with little accumulated precipitation associated with it. On the 10th, the trough is over the Ethiopian Highlands where precipitation accumulation is increasing. There is a large daily accumulated precipitation flux on the 11th as the trough begins to strengthen. The trough continues to mature on the 12th, but the accumulated precipitation decreased. This is consistent with the Gridsat showing weak convection crossing Sudan at 12/06Z and 12/18Z in Figure 3b. By the 13th, close contours of geopotential heights at 600 mb located at (7°N, 19°E) indicate a mature easterly wave organizing on the lee side of the Darfur Mountains. It is noteworthy to see during the daily averages of geopotential that a subtropical high (Arabian High) over the Arabian Peninsula is present. The pressure increases per day for the Arabian in which it will also increase easterly flow on the southern rim of the subtropical anticyclonic zone over southern Arabia and Easter Africa.

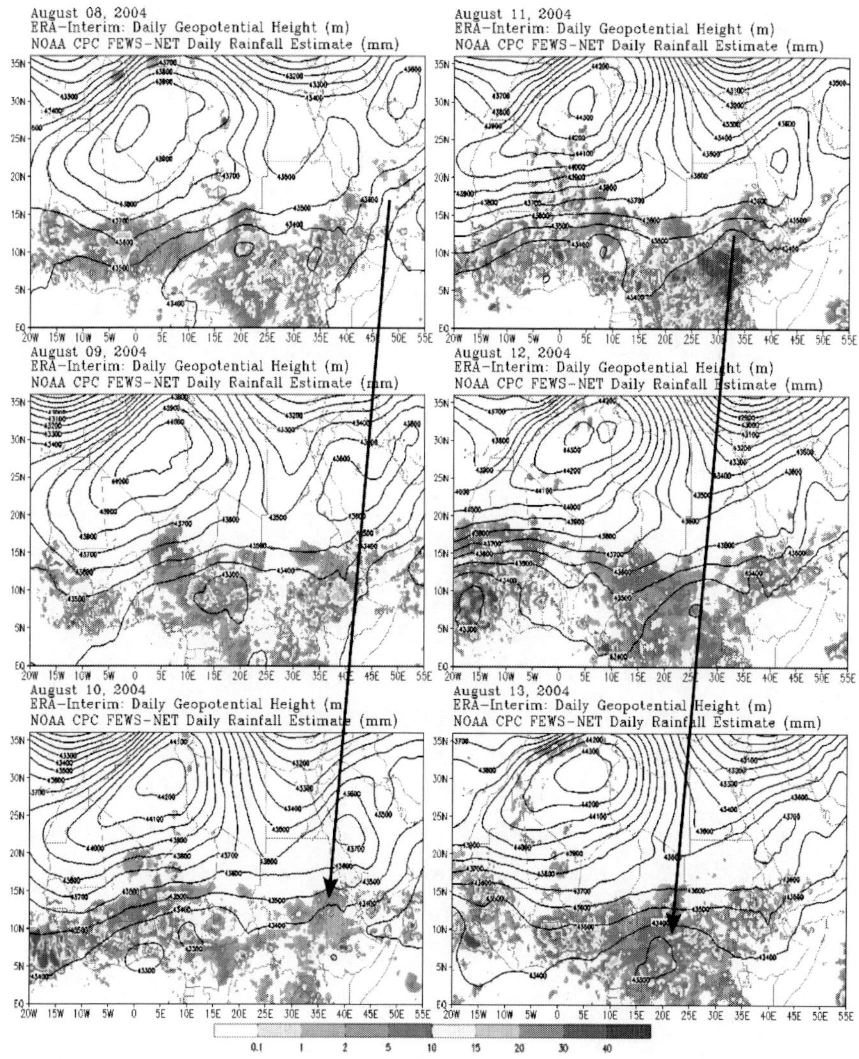

Figure 4. NOAA CPC FEWS daily precipitation (mm) shaded and ERA-I 600 mb daily geopotential heights (m) in black contours between August 8th and August 13th. Black line denotes the location of A04.

3.1.2. Relative Vorticity Analysis Using ERA-I

In section 3.1.1, A04's convective cloud clusters were traced Gridsat and ERA-I total cloud cover. Difficulty with tracking clouds clusters is due to formation and dissipation of clouds with the diurnal cycle. When convection is weak, cloud tops are not as cold according to brightness temperature. This suggests a weak phase for the MCS, but the associated vorticity may still be present for easier tracking. Figure 5 is the 600 mb relative vorticity field for the ERA-I. The band of positive relative vorticity between 5°N and 15°N is the preferred region of easterly wave propagation. Starting on 9/06Z at (14°N, 47°E), there is strong positive vorticity over the southern Arabian Mountains, but the convection is weak according to the GridSat and ERA-I cloud cover analysis at the same time. This vorticity propagates into the East Africa at 10/06Z where it is enhanced by the orographic effects from the Asir Mountains and Ethiopian Highlands. As the vorticity perturbation continue to propagate away from the Ethiopian Highlands at 10/18Z, it begins to strengthen and mature. At 11/18Z, the system is becoming more organized with the convection produced from the Ethiopian Highlands and continues it propagation across Sudan to the Darfur Mountains by 12/18Z. Once the system reaches the Darfur Mountains, it organizes again from orographic effects and matures into an AEW at 13/06Z. At 13/18Z, A04's MCS (Figure 2b) and A04's AEW are now a coupled system and is part of a Rossby wave that extends from West Africa at (15°N, 13°W) to the southern Arabian Peninsula at (12°N, 47°E). From Arabia (47°E) to the lee side of the Darfur Mountains (20°E) is about 2970 km and it took A04's vorticity perturbation 108 h to travel. A04 has a propagation speed of 7.6 m s^{-1}, which is slightly faster than the wave propagation speed observed by (Burpee 1974; Reed et al. 1977; and Chen 2006).

The vorticity budget will be used to explain the generation of A04's vorticity perturbation over the Arabian Peninsula. Using the vorticity equation from (Bluestein 1992) we have:

$$\frac{\partial \zeta}{\partial t} = -\boldsymbol{V} \bullet \nabla\zeta - \omega\frac{\partial \zeta}{\partial p} - v\beta - (\zeta + f)\nabla \bullet \boldsymbol{V} - \hat{k} \bullet \nabla\omega \times \frac{\partial \boldsymbol{V}}{\partial p} + \hat{k} \bullet \nabla \times Fr \quad (1)$$

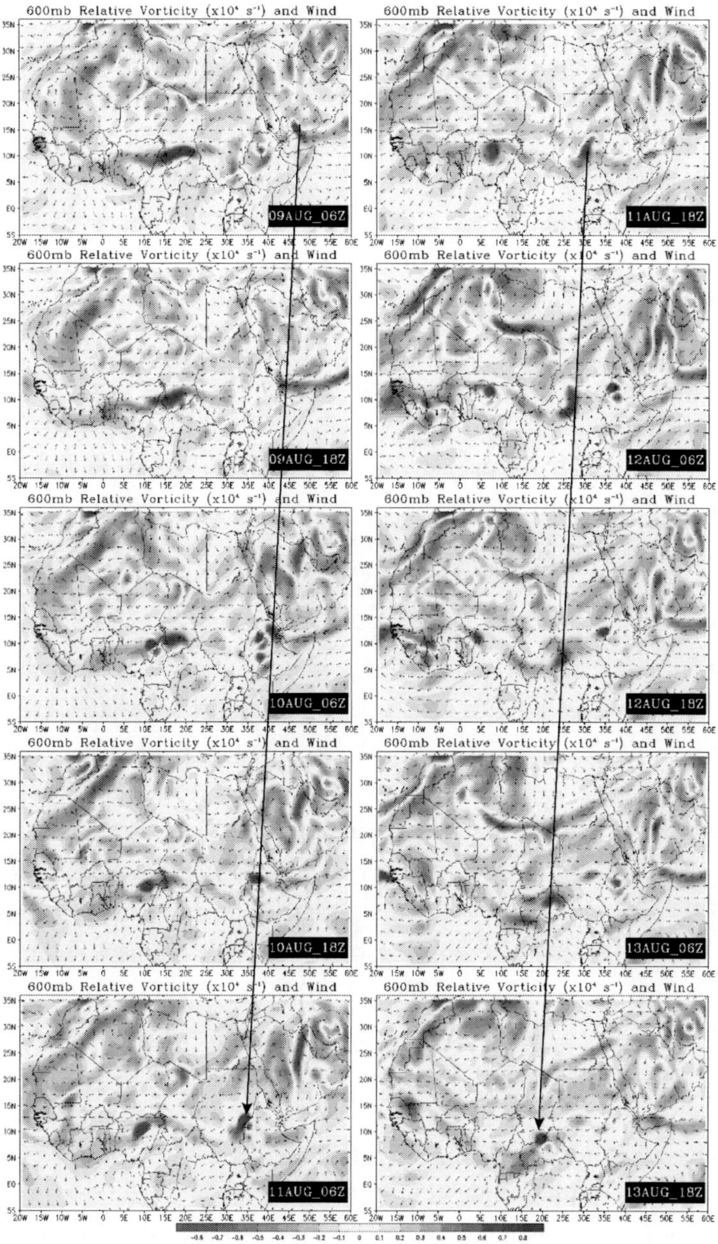

Figure 5. ERA-I 600 mb relative vorticity (s^{-1}) and wind vectors between 08/09/06Z to 08/13/18Z every 12 h. Red (blue) shade is for positive (negative) vorticity. Black line represents the track of A04.

Equation (1) examines vorticity generation from a mechanical perspective. After analysis of the relative vorticity field, earliest detection of A04's vorticity perturbation was at 09/00Z. Magnitudes for local rate of change, horizontal vorticity advection, vertical advection, and divergence are at an order of 10^{-10} s^{-2}. The local rate of change in vorticity, shows the evolution of vorticity between 08/18Z to 09/06Z (Figure 6a). Central differencing in time from the local rate of change yields the detection time of A04's vorticity perturbation located (15°N, 47°E). The positive horizontal vorticity advection has the highest magnitude between all the terms (Figure 6b). Easterly zonal wind flow will aid advection of relative vorticity, therefore adding to the intensity of the magnitude of horizontal advection. The positive vorticity advection of A04 is located at (15°N, 48°E) at 09/00Z. The presence of the Asir Mountains induces vertical motion of vorticity that is generated from the diurnal cycle (Figure 6c). The source of vertical motion is also associated with convection generated from ITCZ heating.

Preexisting relative vorticity (ζ) is in each term of Eq. (1), but the divergence term $-(\zeta + f)\nabla \bullet \boldsymbol{V}$ can be split into two separate terms with ζ and f, respectively. Since f is dependent on the Earth's rotation, then convergence and f will generate vorticity. Figure 6d at 09/00Z, shows A04's vorticity perturbation being generated over the Arabian Peninsula located at (15°N, 48°E). Comparisons were done (not shown) between the magnitudes of $f\nabla \bullet \boldsymbol{V}$ and $\zeta\nabla \bullet \boldsymbol{V}$ and it was concluded that $f\nabla \bullet \boldsymbol{V}$ had the higher magnitude. Given this result, $\zeta\nabla \bullet \boldsymbol{V}$ became negligible in contribution to vorticity generation for A04. The sources of the vorticity perturbations were attributed to convergence of southeasterly Shamal winds, which is associated with the Arabian high and the northwesterly Somali jet winds, that is associated with the Indian Ocean high (Lin et al. 2013). These interaction increases vorticity generation until the vorticity is advected by the mean easterly wind flow of the AEJ.

The beta effect (Figure 6e) shows the drifting planetary vorticity advection. The meridional gradient of planetary vorticity is high over the Arabian Peninsula in comparison to East Africa because of the southerly meridional winds over the Arabian Peninsula are strong.

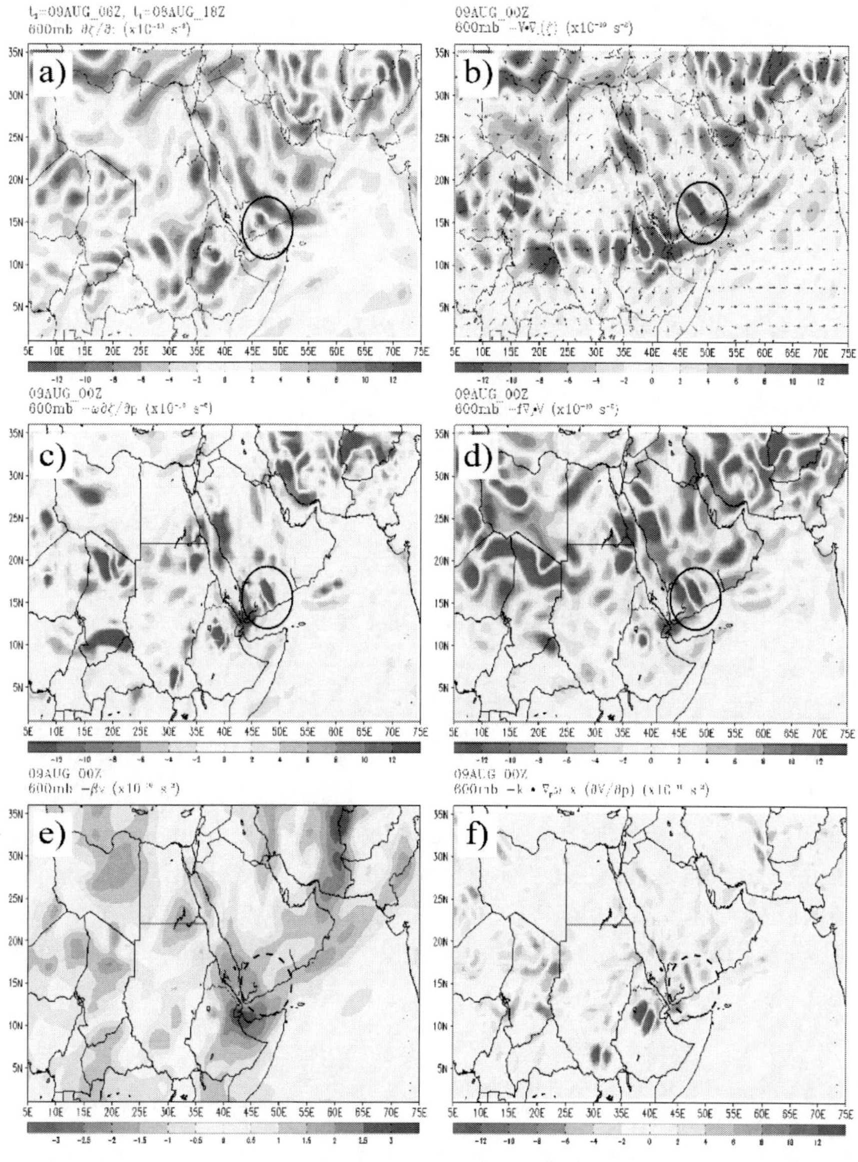

Figure 6. ERA-I 600 mb vorticity budget analysis of (a) local rate of change (x10^{-10} s^{-2}) for t_2 = 09/06Z and t_1 = 08/18Z, (b) vorticity advection (x10^{-10} s^{-2}) at 09/00Z, (c) vertical vorticity advection (x10^{-10} s^{-2}) at 09/00Z, (d) divergence (x10^{-10} s^{-2}), (e) beta effect (x10^{-10} s^{-2}), and (f) tilting (x10^{-11} s^{-2}). The black circles denote where A04 is located.

The southerly winds are associated with the Arabian high and Shamal winds. Even though the magnitudes of the beta term are on the same order of $-\boldsymbol{V} \bullet \nabla\zeta, -\omega\frac{\partial\zeta}{\partial p}, and - f\nabla \bullet \boldsymbol{V}$; the values are smaller than the respective terms. The titling term (Figure 6f) is at an order of 10^{-11} s^{-2}. and therefore, is negligible in comparison with the higher ordered terms.

Table 1. Physics parameterization schemes used to test the performance of the WRF model

Experiment	Microphysics	Cumulus
A.	Thomson	Kain-Fritsch
B.	Thomson	Grell 3d ensemble
C.	WSM–6 Class	Kain-Fritsch
D.	WSM–6 Class	Grell 3d ensemble
E.	Goddard	Kain-Fritsch
F.	Lin	Grell 3d ensemble
G.	Goddard*	Betts-Miller-Janjic*
H.	Lin	Betts-Miller-Janjic

* Combination G is used for the main simulation.

The relative vorticity field and budget analysis of A04's vorticity perturbation using the ERA-I provided realistic details about wave genesis over the Arabian Peninsula. It was found that A04's vorticity perturbation was generated around 09/00Z and was a mature AEW at 13/18Z. This will be essential for setting up a time interval to simulate A04 using the WRF model.

3.2. WRF Simulated Results

3.2.1. Model Performance and Comparison to ERA-I

The WRF simulation uses ERA-I at 08/00Z to initialize and is ran until 15/00Z, outputted every 3 h. A series of experiments were performed to identify possible physics combinations that would best represent A04. Figure 7 illustrates the various combinations using daily rain totals and 600 mb geopotential height with the corresponding combinations in Table 1. Comparing Figure 7 to Figure 4, it can be seen how the model performed against the NOAA CPC FEWS daily precipitation and ERA-I geopotential height. Using precipitation as a diagnostic variable will eliminate physics combinations that may have simulated A04 too aggressively or not aggressive enough.

The most aggressive combinations for precipitation came from WSM 6 Class, Thompson, Kain-Fritsch, and Grell 3d ensemble (Figure 7a-d). The precipitation is a result from convection being over intensified or producing unrealistic convective processes. These results do not compare well with Figure 4. The Goddard and Lin schemes (Figures 7e-h) performed better in terms of precipitation reliability and cumulus schemes were narrowed down to Betts-Miller-Janic (Figure 7g and h). After close inspection and comparison between physic combinations, the Goddard and Betts-Miller-Janic scheme were used as the control simulation for this case study.

The calculated relative vorticity magnitudes are stronger in the WRF simulation in comparison to the ERA-I. This is due to increased grid resolution of 8 km. In Figure 8, the WRF simulated 600 mb horizontal wind (Figure 8a and b) can be compared with the ERA-I data (Figure 8c and d) to show the intensity difference between models. The components of relative vorticity are $\partial u/\partial y$ and $\partial v/\partial x$. An increase of intensity from either component will intensify the overall relative vorticity. The most noticeable difference of intensity comes from the simulated zonal wind (Figure 8a) versus the ERA-I zonal wind (Figure 8c). The easterly wind between 9°N and 21°N illustrates the AEJ. Spinks et al. (2014) investigated the AEJ over East Africa and found a local wind maximum embedded within the AEJ centered at (14°N, 35°E). This local wind maximum in East Africa (LMW_E), is present in the simulation and ERA-I, but the intensity is significantly different. For the WRF simulation and ERA-I, the intensity of the LMW_E is 12 to 14 m s^{-1} and 10 to 12 m s^{-1}, respectively. Westerly wind has a stronger presence over East Africa centered at (6°N, 20°E) in the simulation versus the ERA-I. The magnitude of the easterly wind is stronger over the Arabian Peninsula as well.

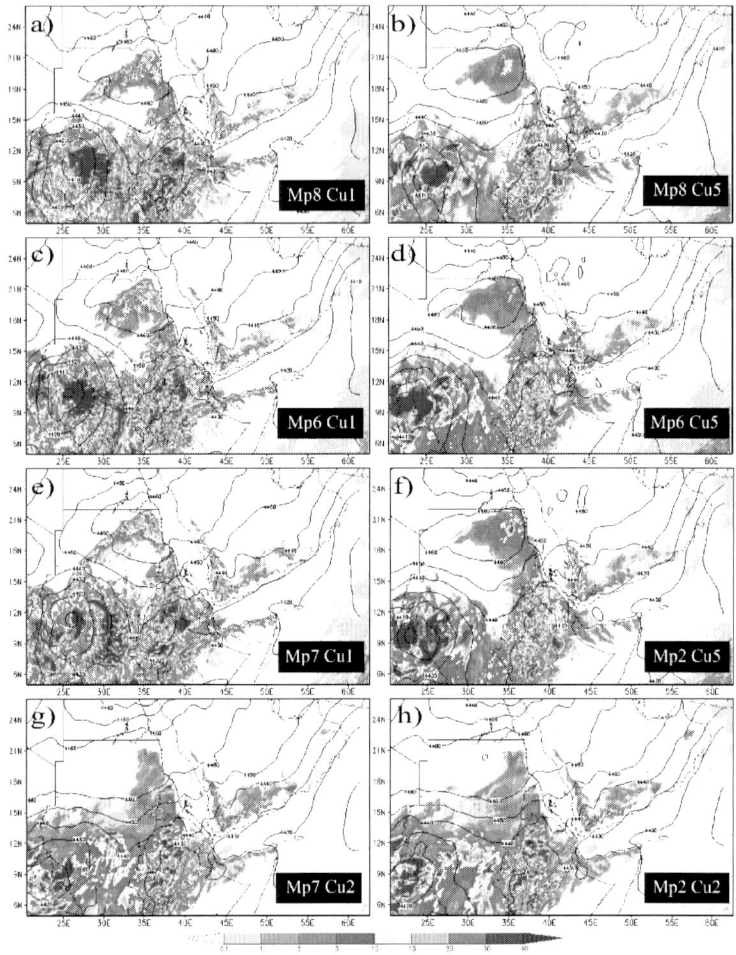

Figure 7. WRF simulated daily precipitation (mm) of August 12, 2004 (shaded) and 600 mb geopotential height (m) in black contours for multiple physics schemes. The corresponding combination of physic schemes can be seen in Table 1.

The reason for the zonal wind magnitude increase from the simulation versus the ERA-I, is due to the geopotential height being stronger. The Arabian high can be seen in the WRF simulation centered at (21°N, 37°E) (Figure 8a) and the ERA-I centered at (21°N, 43°E) (Figure 8c). The geopotential height vales are stronger in the WRF simulation for the Arabian high by about 1000 m and the structure of the anticyclone is broader. The increased geopotential values will, induce a stronger positive meridional geopotential gradient, which will increase the easterly zonal magnitudes that influences relative vorticity intensity over East Africa and the Arabian Peninsula.

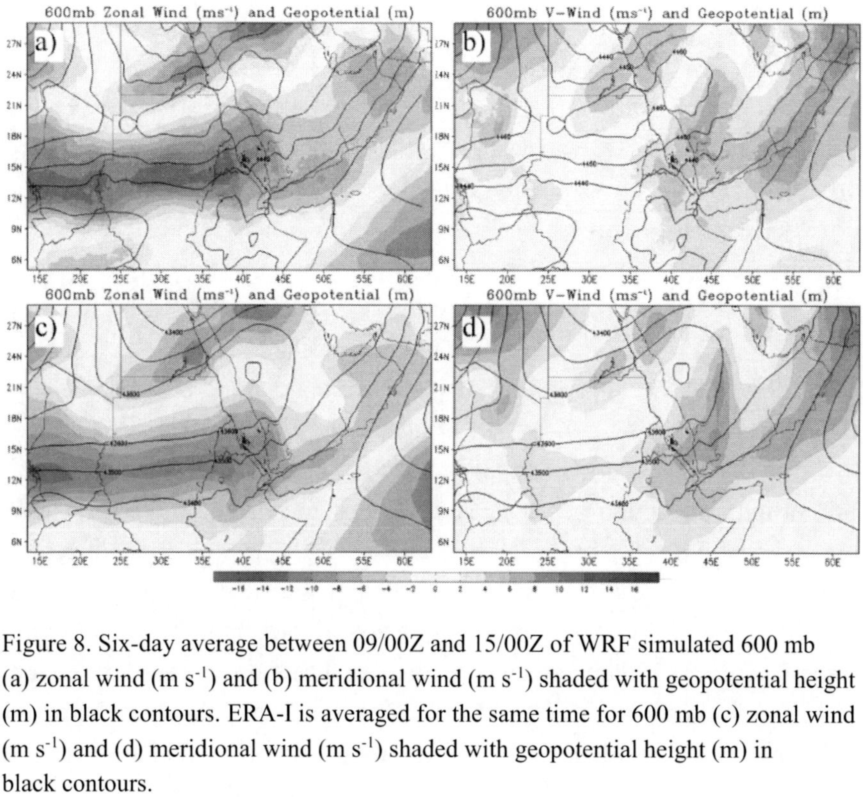

Figure 8. Six-day average between 09/00Z and 15/00Z of WRF simulated 600 mb (a) zonal wind (m s^{-1}) and (b) meridional wind (m s^{-1}) shaded with geopotential height (m) in black contours. ERA-I is averaged for the same time for 600 mb (c) zonal wind (m s^{-1}) and (d) meridional wind (m s^{-1}) shaded with geopotential height (m) in black contours.

The meridional wind comparison between the WRF simulation and ERA-I (Figures 8b and 8d), reveals little difference of magnitudes. This means that the relative vorticity magnitude increase will come from the zonal wind component. The northerly wind over the Arabian Peninsula center at (20°N, 43°E) is associated with the Arabian high. The northerly jet located in southeast Arabian is from the gradient of the subtropical high system and Indian Monsoon low converging.

The relative vorticity magnitudes will be higher in the WRF simulation versus the ERA-I. The WRF simulation will also resolve more vorticity process, providing information that cannot be completely from the ERA-I. The next section will illustrate the WRF simulated vorticity and the 3 stages of case study A04.

3.2.2. Three Stages of A04: Formation, Merger, Mature

Vorticity budget analysis at 600 mb from the WRF simulation will reveal genesis mechanisms that will explain the origin of A04's vorticity perturbation. In section 3.1.2, budget analysis was performed using the ERA-I. In section 3.2.1, it was explained that the calculated relative vorticity in the WRF simulation will have a higher magnitude than the vorticity from the ERA-I. In Figure 9, the vorticity budget from the simulation uses the same variables as Eq. (1) in Figure 6 from the ERA-I. The origin of A04's vorticity perturbation was traced to 9/00Z, so we used central differencing of time for the local rate of change, where t_1 = 08/21Z and t_2 = 09/03Z. The magnitudes of vorticity show an increase of intensity in comparison to the local rate of change in the ERA-I. A04 is located at (15°N, 47°E) and the order of the local rate of change is as high at 10^{-9} s^{-2}. The vorticity advection has the highest order and magnitude between all terms, with Figure 9b being plotted at 10^{-9} s^{-2}. The high magnitudes of vorticity advection can be explained by preexisting vorticity perturbations along the ITCZ. Positive vorticity advection is dominative along the southern Arabian Mountains and will be enhance from orographic effects and mean easterly wind flow in the mid-troposphere. The increased meridional zonal gradients from the relative vorticity calculation is the major component of the enhance magnitude for vorticity advection. It was shown in Figure 8, that the meridional wind component has little increase of magnitude to vorticity, so the meridional vorticity advection component of $-\boldsymbol{V} \bullet \nabla\zeta$ is smaller than the zonal component. Instead, the meridional component is important for cyclonic vorticity generation from to the interactions of the northerly and southerly winds converging at (15°N, 47°E).

The vertical vorticity advection illustrates where the highest magnitude of vorticity will be advected into the atmosphere. For A04, the southern Arabian Mountains was the source of vertical motion induced by the diurnal cycle and easterly wind flow over the mountains. The term $-\omega\frac{\partial\zeta}{\partial p}$ does not explain generation of vorticity, but it does explain the enhancement by vertical advection.

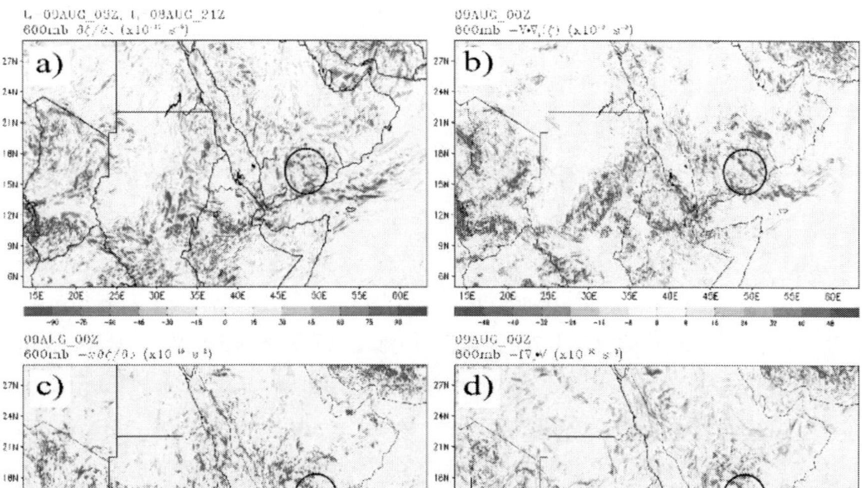

Figure 9. WRF simulated 600 mb vorticity budget analysis of (a) local rate of change ($x10^{-10}$ s^{-1}) for t_2 = 09/06Z and t_1 = 08/18Z, (b) vorticity advection ($x10^{-10}$ s^{-1}) at 09/00Z, (c) vertical vorticity advection ($x10^{-10}$ s^{-1}) at 09/00Z, (d) divergence ($x10^{-10}$ s^{-1}), (e) beta effect ($x10^{-10}$ s^{-1}), and (f) tilting ($x10^{-11}$ s^{-1}). The black circles denote where A04 is located.

The divergent term $-(\zeta + f)\nabla \bullet \boldsymbol{V}$ will be the most important factor to illustrate vorticity generation. Using ζ infers preexisting vorticity and ζ is less than f on a synoptic scale, so it can be negligible for the generation of A04's vorticity perturbation. Thus, we are left with $-f\nabla \bullet \boldsymbol{V}$, which explains the generation of relative vorticity due to convergence. In Figure 9d, A04 is located at (15°N, 48°E) with strong magnitudes of convergence over the southern Arabian Mountains.

The beta term (Figure 9e) and titling term (Figure 9f) has the smallest magnitudes and therefore are not significant for the generation of vorticity. The terms $\frac{\partial \zeta}{\partial t}$, $-\boldsymbol{V} \bullet \nabla\zeta$, and $-f\nabla \bullet \boldsymbol{V}$ are the most significant in magnitude and clearly shows how A04 vorticity is generated from convergence, the advection of vorticity, and the local rate of change of vorticity. The diurnal cycle will help induce orographic generated vorticity along with convergence of the Shamal northerlies and southerlies from the Indian Ocean high. The vorticity advection has the highest order and works complementary with the easterly flow over southern Arabia.

Lin et al. (2005) found that there are two modes of the pre-Alberto (2000) MCSs embedded within the AEW structure: a stationary mode and a propagating mode. The stationary mode exists over the Ethiopian Highland on a diurnal variation due to sensible heating over the high terrain. The propagation mode coincides with the propagation and generation of AEWs in the vicinity of the Ethiopian Highlands. In a later study, Lin et al. (2013) found a stationary mode of over the southern Arabian Mountains, which was the genesis region for the pre-Debby AEW-MCS precursor. The two stationary modes of vorticity generation over the Ethiopian Highlands and Arabian Mountains influence the genesis of A04's vorticity perturbation.

Figure 10a illustrates the formation and evolution phase of A04's associated MCSs and vorticity from the outgoing long wave radiation (OLR) and relative vorticity field, respectively. There are two distinct areas of orographic generated lee vorticity and convection that will contribute to the formation of A04. At 09/06Z, the convective cloud burst and vorticity perturbation at (12°N, 40°E) over the Ethiopian Highlands is denoted as MCS-1.

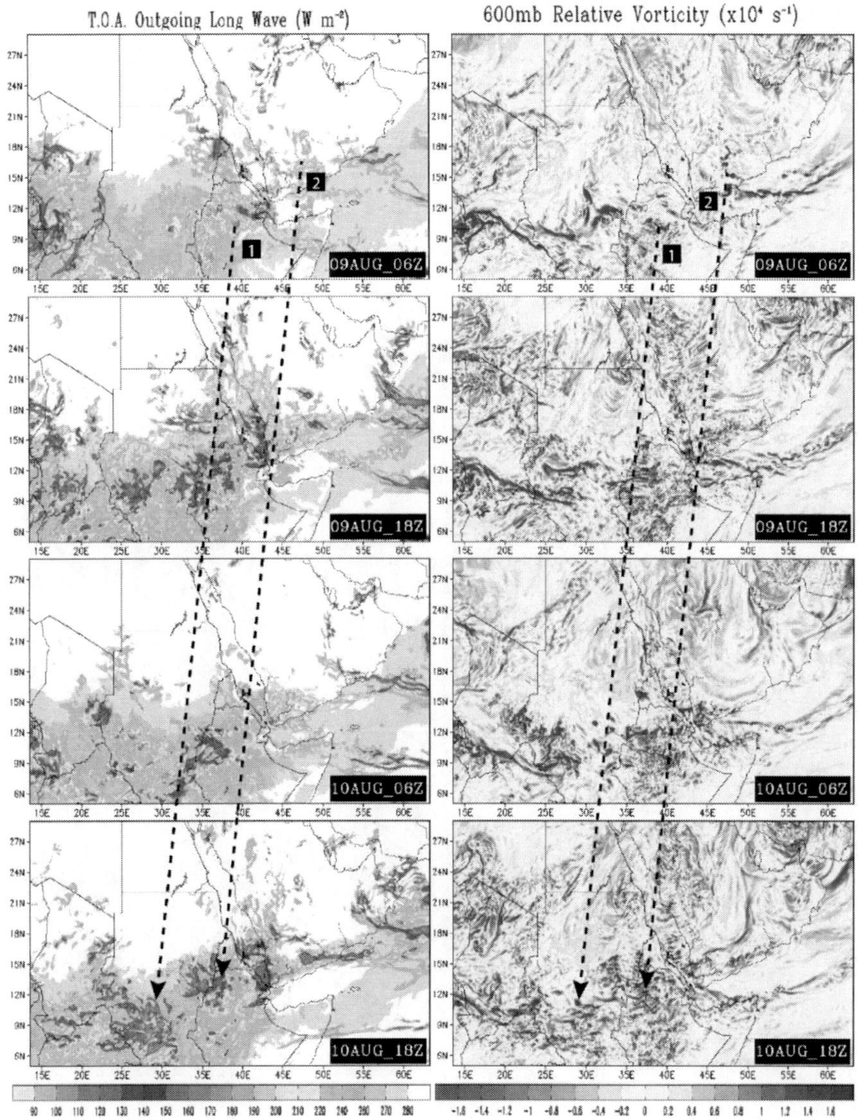

Figure 10. (a) WRF simulated top of the atmosphere outgoing long wave radiation (W s^{-2}) shaded in the left column and 600 mb relative vorticity (10^{-4} s^{-1}) shaded in the right column. Time interval is between /08/09/06Z to 08/10/18Z every 12 h.

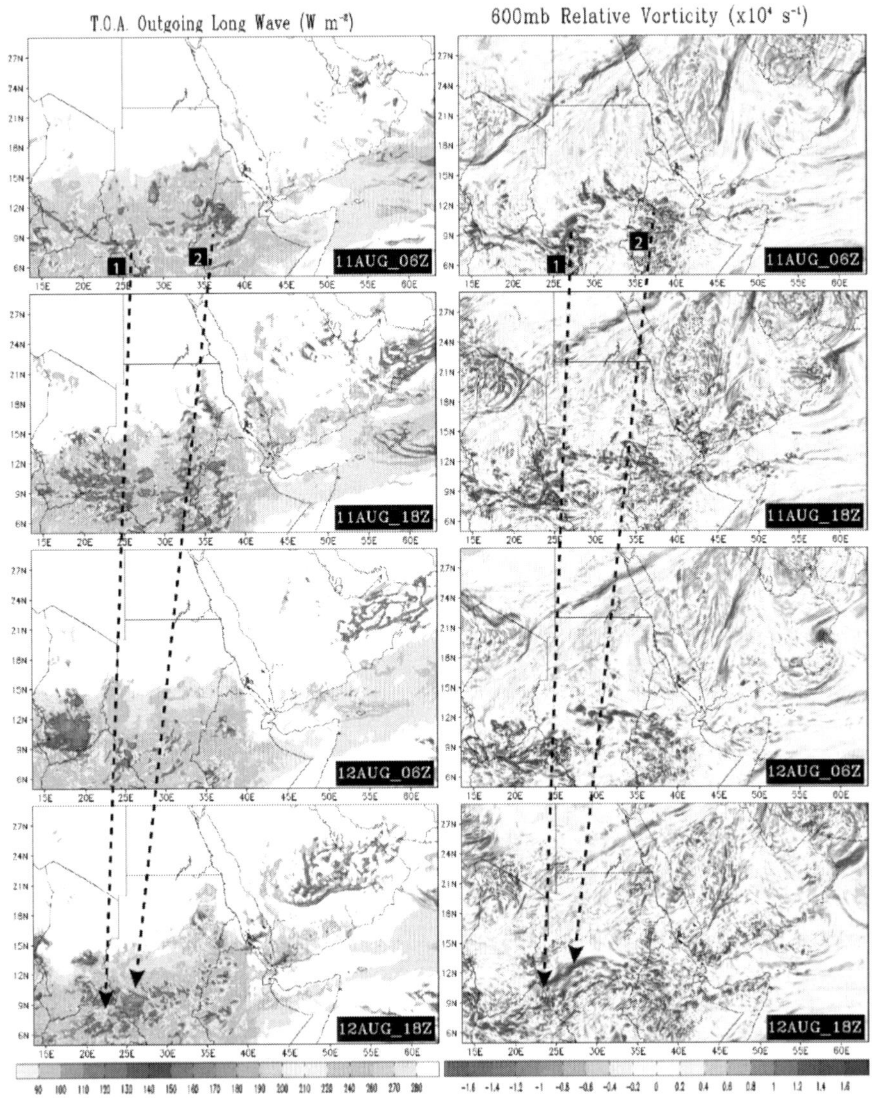

Figure 10. (b) WRF simulated top of the atmosphere outgoing long wave radiation (W s^{-2}) shaded in the left column and 600 mb relative vorticity (10^{-4} s^{-1}) shaded in the right column. Time interval is between /08/11/06Z to 08/12/18Z every 12 h.

The convective cloud burst and vorticity perturbation located at (15°N, 47°E) over the southern Arabian Peninsula is the denoted as MCS-2. These MCSs will propagate to the west where they will be enhanced by the orographic effects. The biggest difference between the relative vorticity fields of the 8-km WRF simulation and the ERA-I is that the simulation resolves more orographic generated vorticities, especially during the diurnal cycles. At 09/18Z, MCS-1 and MCS-2 are being enhanced by the Ethiopians Highlands and Asir Mountains, respectively. The convective cloud burst with MCS-1 becomes better organized, while MCS-2 is propagating into East Africa by 10/06Z. At 10/18Z, the two MCSs are very noticeable by the OLR field. MCS-1 is located over South Sudan at (9°N, 30°E) and MCS-2 is located over the Ethiopian Highlands at (14°N, 37°E). During the formation stage of MCS-1 and MCS-2, the systems were propagating at the same speeds of 8.5 m s^{-1}.

During the merging phase, MCS-1 begins to propagate slower than MCS-2 (Figure 10b). At, 11/06Z MCS-1 is slowing down as it nears the Darfur Mountains while MCS-2 begins to split from the Ethiopian Highlands's stationary mode of convection. Parts of MCS-1 is split by the Darfur Mountains at 11/18Z and a convective burst is seen on the lee of the Darfur Mountains at 12/06Z. Over the next 12 h, remnants of MCS-1 continue to propagate to the west at very slow speeds and MCS-2 begins to catch up. By 12/18Z MCS-1 and MCS-2 begins the merging phase, where the vorticities from both systems being to organize with each other. During the merging phase, MCS-1 has a propagation speed of 3.4 m s^{-1} and MCS-2 has a propagation speed of 8.5 m s^{-1}.

The final stage of the formation of A04 is mature phase (Figure 10c). MCS-1 and 2 are now completely merged together and has become a coupled AEW-MCS. At 13/06Z, A04 can be seen on the lee of the Darfur Mountains as a convectively couple system with vortices. The system propagates off to the west while maintaining as an AEW-MCS until it out of the domain at 14/18Z. The propagation speed of the A04 is 8.5 m s^{-1}.

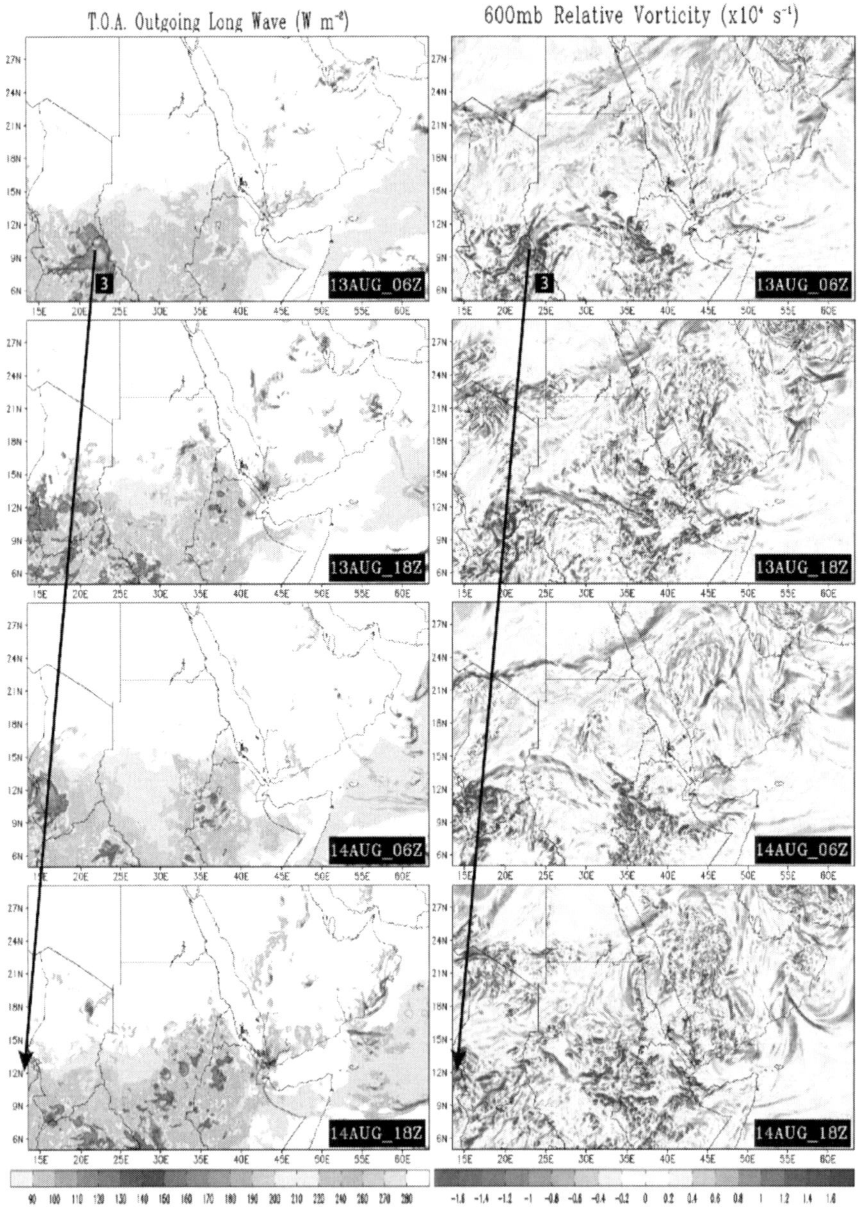

Figure 10. (c) WRF simulated top of the atmosphere outgoing long wave radiation (W s^{-2}) shaded in the left column and 600 mb relative vorticity (10^{-4} s^{-1}) shaded in the right column. Time interval is between /08/13/06Z to 08/14/18Z every 12 h.

3.2.3. Evolution of the AEJ and Embedded East Local Wind Maximum

During A04's evolution across southern Arabia and East Africa, the AEJ played a role in maintenance of the system's intensity. The AEJ have been long understood to support the growth of AEWs through the barotropic/baroclinc mechanism (Charney and Stern 1962; Rennick 1976; Mass 1979; Thorncroft 1995, Grist et al. 2002). In recent studies, it has been found that AEWs may be triggered by MCS over the Darfur Mountains (Berry and Thorncroft 2005; Mekonnen et al. 2006), the Ethiopian Highlands (Lin et al. 2005), and the southern Arabian/Asir Mountains (Lin et al. 2013). This section attempts to show the evolution of the AEJ in East Africa and Arabian Peninsula and the growth of A04 along the jet. The triggering of A04 from MCSs and convergence was already established in the previous section 3.2.2.

The AEJ have two embed local wind maximums (LWM) located in West Africa and East Africa. The intensity of the east local wind maximum (LWM_E) at (35°E, 15°N) has a zonal wind speed average of -9.81 m s^{-1} and the intensity of the west local wind maximum (LWM_W) at (17°W, 18°N) has a zonal wind speed average of -13.41 m s^{-1} for the month of August (Spinks et al. 2014). In particular, the LWM_E is a contributing factor to the maintenance and propagation of easterly waves over southern Arabian and East Africa.

Figure 11 depicts the 600 mb zonal wind and geopotential heights every 24 h. The AEJ is located between 12°N and 21°N and is to the south of the subtropical high-pressure systems. The Arabian high can be seen centered at (25°N, 40°E) with geopotential height values of 4460 m at 9/06Z. Easterly wind over southern Arabia is 10 to 12 m s^{-1} and a maximum located at (15°N, 47°E), which is also associated with MCS-2's vorticity perturbation. In climatological analysis of the AEJ from previous studies, the LWM_E appears to be a stationary feature. In this case study, the LWM_E propagates along with MCS-2 from southern Arabia into East Africa while growing in intensity, reaching its maximum over Sudan. At 10/06Z, the geopotential height increases as well as the LWM_E. The AEJ will be formed at 600 mb with an elongated easterly zonal geostrophic flow generated from baroclinicity. The presence of the Arabian high will increase easterly flow to the south of the anticyclonic centers because of the increased meridional geopotential gradient.

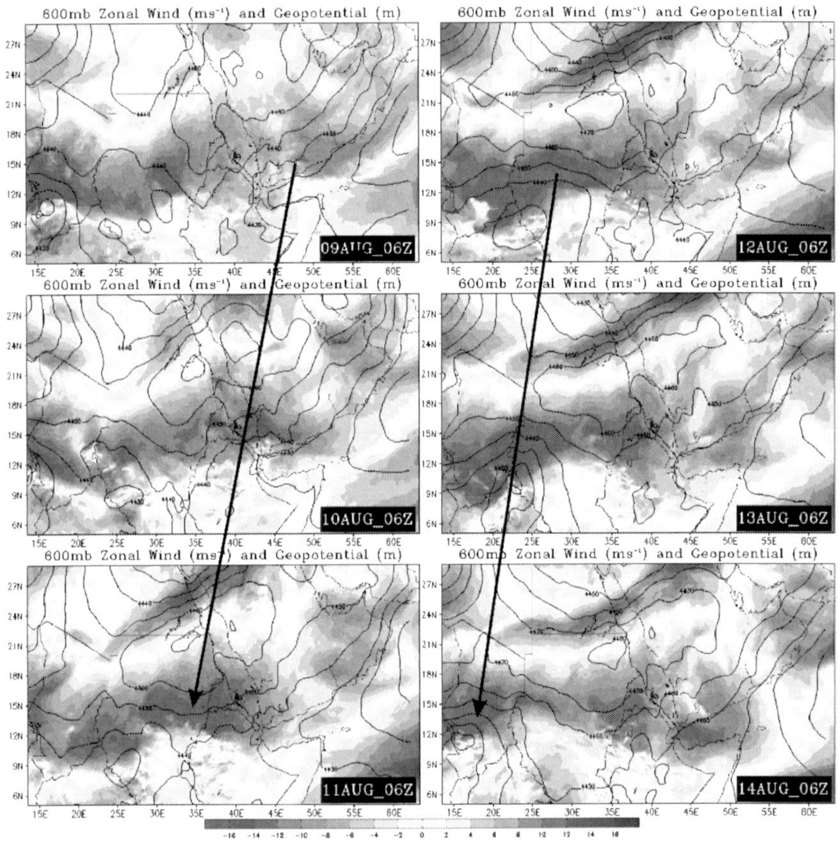

Figure 11. WRF simulated 600 mb zonal wind shaded every 2 m s^{-1} and geopotential height (m) in black contours. Time interval is between 9/06Z and 14/06Z every 24 h. The black lines denote the located of the maximum zonal winds associated with A04.

By 11/06Z, the LWM_E located at (15°N, 35°E) is in a mature phase with velocities greater than 16 m s^{-1} and the Arabian High is centered at (21°N, 35°E). This is also the start of the merging phase of MCS-1 and MCS-2 in section 3.2.2. At 12/06Z, the LWM_E located at (13°N, 27°E) and the presence of westerly wind centered at (5°N, 25°E) will increase the meridional zonal gradient and enhance easterly wave development, which can be seen at (8°N, 24°E). By 13/06Z, the LWM_E is coupled with A04 and is in a mature phase while propagating. As the systems continue to propagate to the west, the LWM_E is separated from the dominate influence of the Arabian high and is intercepted by the Saharan high after crossing 20°E.

The LWM_E is providing maintenance to A04 from the maximum easterly wind flow induced from the Arabian high's meridional geopotential gradient. This component is essential for the survival of vorticity perturbation over the Arabian Peninsula and East Africa. The orographic generated lee vorticities are part of the stationary mode and will need the presence of easterly wind flow in order to propagate away from the mountains. Once split from the mountains, the vorticity perturbations will be able to grow under the influence of the LWM_E. The existence of the westerly wind to the south of the easterly wind will also be a contributing factor to vorticity maintenance when the zonal gradient is robust.

CONCLUSION

A coupled African easterly wave-mesoscale convective system was examined over the Arabian Peninsula and East Africa in August 2004. The system A04 was identified through satellite brightness temperature and ERA-I total cloud cover. A04's convective origins were found to be over the southern Arabian Peninsula and propagated in to East Africa. The vorticities field from the ERA-I was then analyzed to track A04's associated vorticity perturbation. From origin to the mature stage of A04, the propagation speed was 7.6 m s^{-1}. The formation of A04's vorticity perturbation was on 8/9/00Z located at (15°N, 47°E). A vorticity budget was then performed to investigate the formation mechanisms from the ERA-I data. It was found that the horizontal vorticity advection has the highest magnitude and convergence was the driving factor to vorticity generation. While the ERA-I provided significant findings for the formation mechanism using vorticity, it does not resolve well the orographic effects from the southern Arabian Mountains and Ethiopian Highlands.

To test the origin and formation of A04, a numerical simulation using the WRF model was employed using the ERA-I to initialize with. Higher resolution results provided the information needed to explain the origin A04's vorticity perturbation, especially over the mountains. It was found that the higher resolution results produced a stronger zonal wind flow over East Africa and the Arabian Peninsula, which would directly increase the magnitudes of the vorticity field. There were also more vorticity processes resolved over the mountains, especially during diurnal cycles. The vorticity budget was performed on the numerical simulation and the results were as follows:

1. The overall vorticity budget from the numerical simulation increased in magnitude, with the vorticity advection having the highest order of 10^{-9} s^{-2} due to the advection along the ITCZ. A04's vorticity perturbation is highly influenced from easterly wind flow and the zonal gradient that increases vorticity intensity.
2. The convergence of planetary vorticity explains the generation of vorticity over the southern Arabian Peninsula. The Arabian high produces northerly and easterly winds over Arabia and the Indian Ocean high produces southerly and westerly winds. When these highs begin to converge, cyclonic vorticity is generated over the southern Arabian Peninsula, particularly during diurnal cycles when lee vorticities are produced from orographic effects.

A04 went through 3 phases: formation, merging, and mature. Two MCSs were associated with the formation stage, which was over the Ethiopian Highlands (MCS-1) and southern Arabian Mountains (MCS-2) at 9/06Z. These areas are part of the stationary mode of lee vorticity and convective generation from the diurnal cycle. Once generated, the MCSs propagate away from their respected mountain region. During the merging phase starting at 11/06Z, MCS-1 begins to slow down as it nears the Darfur Mountains while MCS-2 continues to propagate at a constant speed of 8.5

m s^{-1} over the Ethiopian Highlands and into Sudan. At 13/06Z, the systems have merged together over the Darfur Mountains and formed into a mature AEW-MCS. The diurnal cycle and convective cloud burst over the next 12 h enhances the systems as it continues to propagate to the west.

The east local wind maximum (LWM_E) of the Africa easterly jet was also investigated to show how A04 is maintained by easterly wind flow. As A04 propagates to the west, the easterly wind flow also increases under the influence of the Arabian high. The high-pressure system induces a strong meridional geopotential gradient that will increase easterly wind on the southern rim of the high. Once over Sudan, the LWM_E is mature and the presence of westerly wind south of the easterly wind flow will produce a stronger zonal gradient. Over Sudan is when MCS-1 and MCS-2 began to merge. Once A04 is mature and propagate to the west of the Darfur Mountains, the LWM_E will be intercepted by the influence of the Saharan high. The data analysis and numerical model simulated results are consistent with the conceptual model proposed in Lin et al. (2013).

ACKNOWLEDGMENTS

We would like to thank J. Zhang, PhD, A. Mekonnen, PhD, and L. Liu, PhD of the North Carolina A&T State University and Dr. J. Lee, PhD of the NOAA ESRL for their comments on this paper. This research was partially supported by the NOAA Educational Partnership Program under the cooperative agreement A06OAR4810187 and NSF Awards AGS-1265783, CNS-1126543, and CNS-1429464.

REFERENCES

Berry, G. J. & Thorncroft, C. D. (2005). Case study of an intense African easterly wave. *Mon Weather Rev.*, *133*, 752-766.

Burpee, R. W. (1972). The origin and structure of easterly waves in the lower troposphere of North Africa. *J. Atmos. Sci.*, *29*, 77-90.

Burpee, R. W. (1974). Characteristics of the North African easterly waves during the summers of 1968 and 1969. *J. Atmos. Sci.*, *31*, 1556-1570.

Carlson, T. N. (1969). Synoptic histories of three African disturbances that developed into Atlantic hurricanes. *Mon. Wea.* Rev., *97*, 256-276.

Charney, J. G. & Stern, M. E. (1962). On the stability of internal baroclinic jets in a rotating atmosphere. *J. Atmos. Sci.*, *19*, 159-172.

Chang, C. B. (1993). Impact of desert environment on the genesis of African wave disturbances. *J. Atmos. Sci.*, *50*, 2137-2145.

Chen, T. C. (2006). Characteristics of African easterly waves depicted by ECMWF Reanalysis for 1991-2000. *Mon. Wea. Rev.*, *134*, 3539-3566.

Dee, D. P. et al. (2011). The ERA-Interim reanalysis: configuration and performance of the data assimilation system. *Quart. J. Roy. Meteor. Soc.*, *137*, 553-597.

Fink, A. H. & Reiner, A. (2003). Spatio-temporal variability of the relation between African easterly waves and West African squall lines in 1998 and 1999. *J. Geophys. Rev.*, *108* (D11), 4332.

Glickman, T. S. (ed.) (2000). Glossary of Meteorology. 2nd Ed., *Amer. Meteor. Soc.*, 855pp.

Grist, J. P., Nicholson, S. E. & Barcilon, A. L. (2002). Easterly wave over Africa. Part II: Observed and modeled contrast between wet and dry years. *Mon. Wea. Rev.*, *130*, 212-225.

Hodges, K. I. & Thorncroft, C. D. (1997). Distribution and statistics of African mesoscale convective weather systems based on ISCCP Meteosat imagery. *Mon. Wea. Rev.*, *125*, 2821-2837.

Hsieh, J. S. & Cook, K. H. (2005). Generation of African easterly wave disturbances, Relationship to the African easterly jet. *Mon. Wea. Rev.*, *133*, 1311-1327.

Hsieh, J.-S. & Cook, K. H. (2007). A study of the energetics of African easterly waves using a regional climate model. *J. Atmos. Sci.*, *64*, 2212-2230.

Knapp, K. R., Ansari, S., Bain, C. L., Bourassa, M. A., Dickinson, M. J., Funk, C., Helms, C. N., Hennon, C. C., Holmes, C. D., Huffman, G. J., Kossin, J. P., Lee, H.-T., Loew, A. & Magnusdottir, G. (2011).

Globally gridded satellite (GridSat) observations for climate studies. *Bull. Amer. Meteor. Soc.*, *92*, 893-907.

Kwon, H. J. (1989). A re-examination of the genesis of African waves. *J. Atmos. Sci.*, *46*, 3621-3631.

Laing, A. G. & Fritsch, J. M. (1993). Mesoscale convective complexes in Africa. *Mon. Wea. Rev.*, *121*, 2254-2263.

Laing, A. G., Trier, S. B. & David, C. A. (2012). Numerical simulation of episodes of organized convection in tropical northern Africa. *Mon. Wea. Rev.*, *140*, 2874-2886.

Landsea, C. W. (1993). A climatology of intense (or major) Atlantic hurricanes. *Mon. Wea. Rev.*, *121*, 1703-1713.

Lin, Y. L, Robertson, K. E. & Hill, C. M. (2005). Origin and propagation of a disturbance associated with an African easterly wave as a precursor of Hurricane Alberto (2000). *Mon. Wea. Rev.*, *133*, 3276-3298.

Lin, Y. L. (2007). *Mesoscale Dynamics.* Cambridge University Press, 630 pp.

Lin, Y.-L, Liu, L., Tang, G., Spinks, J. & Jones, W. (2013). Origin of the pre-Tropical Strom Debby (2006) African easterly wave-mesoscale convective system. *Meteor. Atmos. Phys.*, *120* (1-2), 123-144.

Mass, C. (1979). A linear primitive equation model of African wave disturbances. *J. Atmos. Sci.*, *36*, 2075-2092.

Mekonnen, A., Thorncroft, C. D. & Aiyyer, A. R. (2006). Analysis of convection and its association with African easterly waves. *J. Climate*, *19*, 5405-5421.

Mozer, J. B. & Zehnder, J. A. (1995). Lee vorticity production by large-scale tropical mountain ranges. part ii: a mechanism for the production of African waves. *J. Atmos. Sci.*, *53*, 539-549.

Payne, S. W. & McGarry, M. M. (1977). The relationship of satellite inferred convective activity to easterly waves over West Africa and the adjacent ocean during phase III of GATE. *Mon. Wea. Rev.*, *10*, 413-420.

Pielke, Jr., R. A. & Landsea, C. W. (1998). Normalized hurricane damages in the United States: 1925-95. *Weather Forecasting*, *13*, 621-631.

Rao, P. G., Hatwar, H. R., Al-Sulaiti, H. H. & Al-Mulia, A. H. (2003). Summer Shamals over the Arabian Gulf. *Weather*, *58*, 472-478.

Reed, R. J., Norquist, D. C. & Recker, E. E. (1977). The structure and properties of African wave disturbances as observed during Phase III of GATE. *Mon. Wea. Rev.*, *105*, 317-333.

Rennick, M. A. (1976). The generation of African waves. *J. Atmos. Sci.*, *33*, 1955-1969.

Simmons, A. J. (1977). A note on the instability of the African easterly jet. *J. Atmos. Sci.*, *34*, 1670-1674.

Skamarock, W. C., Klemp, J. B., Dudhia, J. D., Gill, O., Barker, D. M., Duda, M. G., Huang, X.-Y., Wang, W. & Powers, J. G. (2008). A description of the advanced research WRF version 3. *NCAR technical note*, 113 pp.

Thorncroft, C. D. & Hoskins, B. J. (1994a). An idealized study of African easterly waves. Part I: a linear view. *Quart. J. Roy. Meteor. Soc.*, *120*, 953-982.

Thorncroft, C. D. & Hoskins, B. J. (1994b). An idealized study of African easterly waves. Part II: a nonlinear view. *Quart. J. Roy. Meteor. Soc.*, *120*, 983-1015.

Thorncroft, C. D. (1995). An idealized study of African easterly waves. Part III: more realistic basic states. *Quart. J. Roy. Meteor. Soc.*, *121*, 1589-1614.

Thorncroft, C. D., Hall, N. M. & Kiladis, G. N. (2008). Three-dimensional structure and dynamics of African easterly waves. Part III: genesis. *J. Atmos. Sci.*, *65*, 3596-3607.

In: North Africa
Editor: Stephen Fuller
ISBN: 978-1-53612-983-0

Chapter 2

THE PROSPECT OF THE SAHARA SOLAR ENERGY RESEARCH CENTRE IN ALGERIA AS THE FIRST STAGE OF THE SAHARA SOLAR BREEDER INITIATIVE AND A PROPOSAL FOR THE SECOND STAGE

A. Boudghene Stambouli[1,2,*], M. T. Benmessaoud[1], S. Flazi[1], S. Yamaguchi[3, 2], K. Itaka[4], M. Sumiya[5] and H. Koinuma[2,6]

[1]Electrical and Electronics Engineering Faculty
University of Sciences and Technology of Oran, USTO-MB, Algeria
EL M'Naouer, Oran, Algeria
[2]Sahara Solar Breeder Foundation International, Algeria
[3]Center of Applied Superconductivity and Sustainable Energy Research (CASER), Chubu, University, Kasugai, Aichi, Japan
[4]NJRISE, Hirosaki University, Aomori, Japan
[5]National Institute for Materials Science, Tsukuba, Japan
[6]Graduate School of Frontier Sciences, Tokyo University, Japan

* Corresponding Author: Tel/Fax: 0021341627163; E-mail: stambouli@ssb-foundation.com, aboudghenes@gmail.com.

ABSTRACT

Global energy consumption is projected to increase, even in the face of substantial declines in fossil energy intensity, at least 2-fold by mid-century relative to the present because of population and economic growth. Currently the world's growing thirst for oil amounts to almost 1000 barrels a second. The ever increasing world energy demand cannot be satisfied much longer with fossil fuels. Achieving solutions to energy problem that we face today requires long-term potential actions for sustainable development. In this regard, renewable energy (RE) resources appear to be crucial and the one of the most efficient and effective solutions for sustainable development. That is why a shift to a more sustainable mix by means of the use of RE sources, an increase of the energy efficiency, energy sobriety and greenhouse gas mitigation measures is needed. The development of technological innovation in the energy fields is central to the success of the energy paradigm shift worldwide. Among RE resources, solar energy is by far the largest untapped resource and major primary energy exploitable resource. Algeria has the highest average solar radiation per square meter of any country in the world and is an interesting case study in that it had quite a slow start in this field but is now forging ahead with RE –at least at the utilities level, boosted by government policy. Progress at residential level is not as fast, though. This has to be rectified for the country to meet the targets set in its National Development Plan. A broad range of intensive and promising research and development is being conducted, under the Sahara Solar Breeder (SSB) project concept, in order to produce technological options that can allow energy security, climate stabilisation and economic development from solar energy. In view of the intermittency of insolation, SSB project provides an especially attractive approach in which solar energy must be dispatched on demand to the end user using High Critical Temperature Supra Conductor (HT_cSC) cables. The magnitude of such implied infrastructure transition suggests the need for massive investments in innovative energy research. Anticipated patterns of future energy use and consequent environmental impacts are comprehensively discussed in this article. Also, potential solutions to current problems are identified along with the adoption of the SSB project plan and activities. The relations between RE and sustainable development are described with practical cases of SSB energy policy and its portfolio directed towards continental and global clean energy supply with existing solar grade Silicon production from Sahara sand and diatoms technologies making photovoltaic (PV) a significant contributor within a portfolio of energy sources in the coming 10 to 50 years. An illustrative example of some developed SSB key innovative technologies is presented and more should be challenged together, Algerian and

Japanese SSB research teams, to seek the paradigm shift in global energy production and transport system. In both cases, the research is emanating from the University of Sciences and Technology of Oran (USTO-MB) in line with Saida University.

Keywords: silicon, Sahara sand, diatoms, photovoltaic, superconducting global electric grids

LIST OF ABBREVIATIONS

APRUE	National Agency for the Promotion and Rationalisation of Energy Use
ANDI	National Agency of Investment Development, National Fund for Energy Efficiency
NFEE	CREDEG Research and Development Centre of Electricity and Gas
CRTSE	Research Centre in semi-conductor Technology for the Energetic
CDER	Renewable Energy Development Centre
FNRS	Fund for Scientific Research
FNME	National Fund for energy conservation
CREAD	Centre for Research in Applied Economics for Development

INTRODUCTION

Energy security concerns along with consumption growth are rapidly rising in importance in almost all countries. Abundant and economical energy is the life blood of modern civilizations. According to the estimation done by the International Energy Agency (IEA), a 53% increase in global energy consumption is foreseen by 2030; energy security is becoming a serious issue as fossil fuels are non-RE and will deplete eventually in the near future. Moreover, the energy generates from the combustion of fossil fuels has simultaneously created several environmental concerns which can threaten the sustainability of our ecosystem. One of the primary concerns will be the emissions of greenhouse gases and other types of air pollutants such as hydrocarbons, nitrogen oxide and volatile organic compounds. In addition, the world's oil resources will peak within a few decades to come (80 years at the most as depicted in Figure 1) (ExxonMobil, 2012).

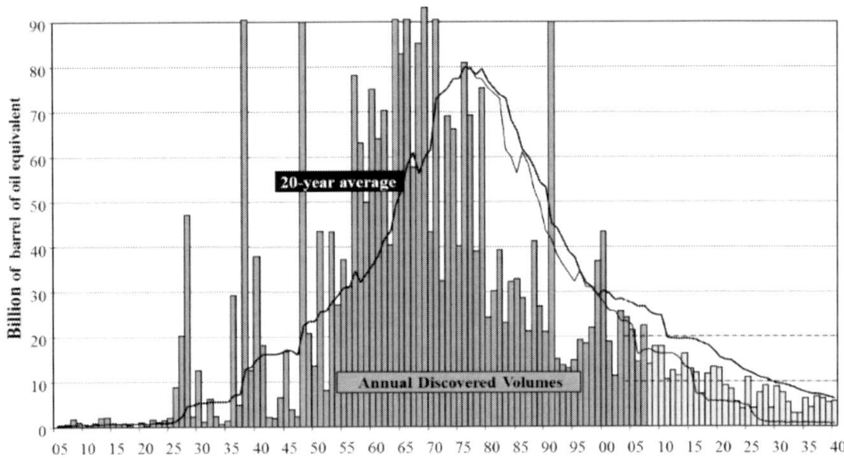

Figure 1. Confronting the supply challenges, discovered annual volume of oil (1905–2040).

Figure 2. Elements for sustainable development.

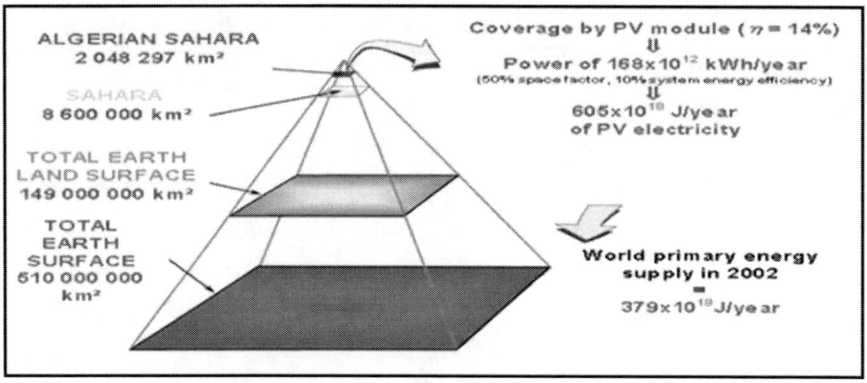

Figure 3. Solar pyramid, the Algerian desert, dead space can be a treasure island of energy used by humanity.

Nuclear power stations are questionable as a main energy source since their vulnerability and serious hazards were realised following the Fukushima Daiichi nuclear disaster on March 11, 2011. In the search for other sustainable alternatives to mitigate some political, economic and environmental issues currently associated with the heavy reliance on fossil fuel, it is inevitable that the world is heading towards RE and new energy economy to reach sustainability (Figure 2) by promoting clean energy technologies, pursuing energy efficiency and developing RE forms which are three orders of magnitude larger than current global energy use (A. Boudghene Stambouli and H. Koinuma, 2012).

All REs, other than geothermal and tidal, derive their energy from the sun being an abundant energy resource. Solar energy is the most promising source of clean, renewable energy, and it has the greatest potential of any power source to solve the world's energy problems (M. Z. A. A. Kadir and Y. A. Rafeeu, 2010). Although solar power producing devices have been around for over 50 years, solar electricity devices often referred to as PV are still considered as cutting edge technology. Globally, there are about 1700 TW of solar power are theoretically available over land for PV. The capture of even 1% of this power would supply more than the world's power needs. Algeria has the potential to be one of the major contributors in solar energy and become a role model to other countries in the world as shown in Figure 3 (A. Boudghene Stambouli et al., 2012) The implementation of various policies and programs by the government of Algeria has increased the awareness of the importance of the role of RE in a sustainable energy system.

In this contribution a particular attention is being given to the joint event that bring together the relevant parties, six Japanese Universities, USTO-MB, Japan International Corporation Agency (JICA), Japan Science and Technology Agency (JST) to develop a long-term vision and strategy to boost the ideas for the realisation and the development of the Sahara Solar Energy Research Centre (SSERC) under the framework of SSB project (A. Boudghene Stambouli et al., 2014, A. Boudghene Stambouli et al. 2015 and H. Koinuma et al., 2014). The SSB plan starts from basic material research on innovative solar Si technology (more than and beyond Siemens processes) and HT_cSC for DC transmission S. (Yamaguchi et al., 2015). This project will also tackle the key challenges and issues related to the field of PV material and systems putting forward the USTO-MB perspective and promoting its R/D activities by a collaborative research plan between Japan and Algeria to identify the most important challenges facing both the research and economic sectors. Although various semiconductors are available for PV, we can deduce the sole feedstock to enable 100 GW annual solar cell manufacture is silicon as judged from total volume of natural resources. In addition to the well-recognised two values of vast land and sunshine of Algeria, its Sahara desert has the third value, sand and Diatoms, for their main component, SiO_2. Subsequently, Algeria can then become a role model to other countries in the world that has huge solar and Si raw material feedstock. By analysing scientific (thermodynamic and kinetic) and technological problems of conventional Siemens Si process, we have designed two strategies to satisfy these requirements: More than Siemens and Beyond Siemens. The latter is aimed extra cost-down of Si by the purification of SiO_2 and C to minimize B and P levels prior to the direct reduction of SiO_2 with C or H and subsequent solidification for removing metallic impurities, thus materialising the third value of desert by converting the sand's primary component silica, the most abundant natural resource on the earth, into solar Si (H. Koinuma et al., 2011). Promising research and work are being done on Diatoms which can be purified and reduced more easily at even lower energy consumption and that could boost advancement on Si reduction process (M. Zerdali et al., 2015). In order to speed up the electric superhighway, the generated electricity from SSB, made up of a network of Very Large Scale-PV (VLS-PV) Power Systems in Sahara desert, will have to be transferred to the North Africa, then Europe, Africa, and ultimately the rest of the world, via HT_cSC that can provide, in compact dimension, firm capacity for base load, intermediate and peaking power, effectively complementing conventional electricity sources (A. Boudghene Stambouli, S. Flazi et al., 2014). The initiative research directed towards Sahara Solar Breeder (SSB) plan (A. Boudghene Stambouli et al, 2012) started in 2010 as the joint program between Japan and Algeria to attract interest and support of neighbouring countries (ssb-foundation) to reach a goal of SSB Initiative in 2015 and hope could extend this basic joint research to industrial cooperation and business in the near future.

SSB Plan as Stem Technology Initiative for Future Energy and Environment Paradigm

In our daily life, energy is present under three main forms: Electricity for lighting, heating and powering appliances; Fuels for mobility and large industrial equipment and Molecules for the ubiquitous chemical compounds making our modern life possible. The development of technological innovation in these fields is central to the success of the energy paradigm shift. The energy transition trajectory is country-specific as it depends on the local available resources, the system in place (energy mix, infrastructure, typology of consumption…) and the societal context. The future energy and environment paradigm is the shift to transform the present energy system mainly based on depleting commodities into a more sustainable mix by means of the use of RE sources, an increase of the energy efficiency, energy sobriety and greenhouse gas mitigation measures.

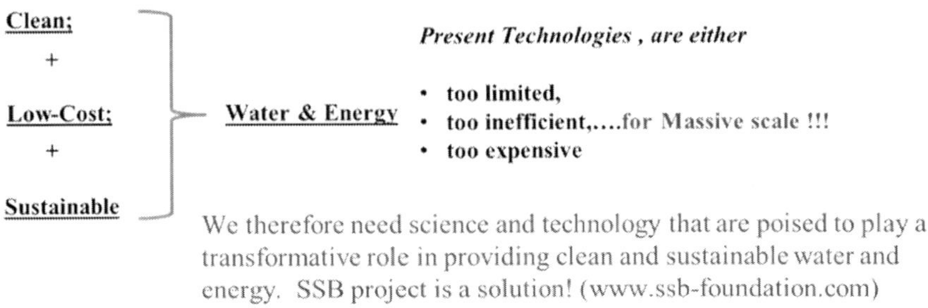

Figure 4. Emerging issues need to be addressed in the society.

SSB may be a great solution towards world sustainability since it proposes a plan to overcome issues the society is facing today as shown in Figure 4 (A. Boudghene Stambouli et al., 2015) and summarised as follows:

- Climate change (mostly from burning of fossil fuel).
- Increase in the prices of limited energy resources
- Increase in the prices of energy-intensive commodities
- Air and water pollution
- Economic development in all aspects
- National and international security
- Poverty and global health

The theoretical potential of solar energy is estimated at 10.8×10^8TWh per year, but producing this energy would require the full use of all available land area, at 100% conversion efficiency. Determining the technical potential requires assessing the fraction of land that can practically be used as well as realistic conversion efficiency. Estimates for this quantity range from 0.44×10^6TWh (1580 EJ) per year to 1.4×10^6TWh (5122 EJ) per year—that is, from 3.1 to 10.2 times the world's primary energy consumption rate in 2007 (BP Statistical Review of World Energy, 2014). Figure 5 shows how we need science and technology that are poised to play a transformative role in providing clean and sustainable energy (A. Boudghene Stambouli et al., 2015). The SSB project may be a solution since it proposes a plan of international partnership in basic research and development, industrial production, trade and finance to construct, gradually over the coming decades, a 'Global Clean Energy Superhighway' starting in the Sahara desert in North Africa beginning from Algeria.

Recognising that the desert is filled with sand, the most abundant natural resource on the earth, and that the sand could be the third value of desert by chemically converting it into solar silicon, the SSB, plan proposed by the Science Council of Japan (H. Koinuma et al., 2009), is one option that is attracting much MENA attention, partly because of the possibility of using desert sands for PV power generation to supply the energy needs of the world. There are two renowned big benefits of Sahara and other deserts from the viewpoint of energy utilisation, i.e., vast land and ample sun shine, which attract much interest for harvesting new clean energy from deserts. Therefore, Material and energy are key components of our life and indispensable for sustainable development of our world.

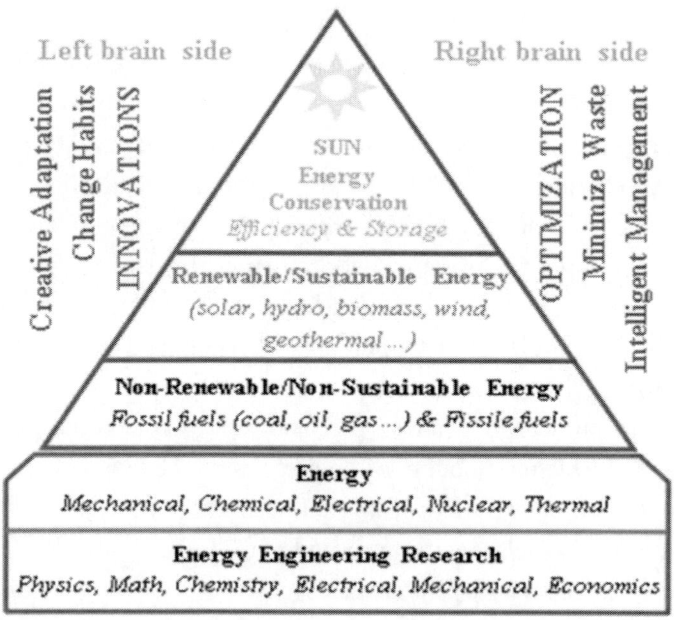

Figure 5. Thrust areas of energy engineering, forms of energy and energy pyramid.

Sahara desert is mostly empty and receives about as much reliable sunlight as you could possibly hope for, but what really makes it a great place to construct a bunch of solar power plants is that the Sahara is completely buried in the raw materials for constructing solar cells: sand.

SSB, is a project which proposes a plan of international partnership in basic research and development, industrial production, trade, financing, etc., to construct, gradually over the coming decades, a Global Clean Energy Superhighway starting in the Sahara desert in North Africa (beginning from Algeria). SSB will help migrate from the unsustainable current excessively fossil-fuel-based global energy paradigm to a more sustainable one. It will also help meet global energy challenges, and mitigate climate change and other environmental problems. SSB is more than an energy solution. It is an integrated community, socio-economic, industrial, agricultural, environmental, and science and technology development solution. In particular, through desalination and adequate irrigation technologies, it will help mainstream marginal desert water resources and lands back into national, regional and international development processes.

In August the 4th 2010, USTO-MB and JICA/JST formed a joint group and signed a series of Memorandum of Understanding and bilateral agreements to announce the creation of a Sahara Solar Energy Research Centre (SSERC), at the USTO-MB, to pursue the realisation and the development of PV plant at a chosen experimental site at the University of Saida in the south west of Algeria also known as the gate of the Algerian Sahara. The SSB plan involves building manufacturing plants around the Sahara desert that would extract silica from the sand and turn it into solar panels to generate RE. The RE from the first facility will then be used to breed more manufacturing plants and, in turn, more solar panels to generate ever increasing amounts of solar power. The ultimate goal is to build enough plants until the breeding strategy can deliver enough green energy of 100 GW of electricity to provide 50% of the world's electrical power generation capacity by 2050 which would be delivered via a global superconducting electrical grid to turn the world's biggest desert into the world's biggest power station taking advantage of two resources that are found in abundance in the Sahara namely silica and sunlight (Figure 6) (K. Kurokawa et al., 2009).

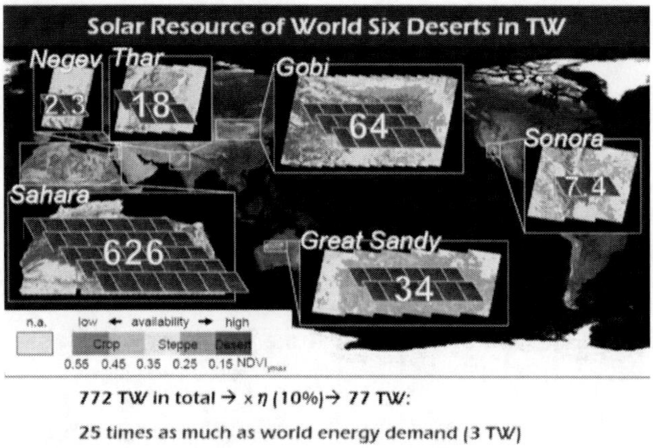

Figure 6. Solar resources of the world six deserts (TW).

The abundant sunlight in Algeria makes it good place for solar power generation and plenty of sand and Diatoms from which to extract silicon of more than 99.9999% purity. Sand in the Sahara is well-suited for that because the sand has high silica content and fewer impurities than sand from other places. The researchers hope to mass-produce the key material used in solar cells at low cost. The SSB perspective of investing in the Solar Energy gives rise to unfolded benefits. Developing solar PV energy will not only establish ground to deal with contemporary conflict issues such as scarcity of water and energy. It will also give the desert a value through its potentially vast possession of power and water sources. This is also supplemented with transfer of know-how and employment to people for the construction VLS-PV power plants and HT_cSC transmission lines.

SSB concept is a role model for the advanced world as depicted in Figure 7. It focuses on a holistic approach to generate two fundamental needs; energy and silicon from unused lands in the desert.

The project apart from being eco-friendly is also economically rewarding based on various analyses of cost and revenues (Corporate Energy Solutions, Environmental Study, Solar Energy, PV Solar Generators, PV Solar Power Plants…). SSB is primarily distinguished by the PV power generation and by the aims to explore further values of desert as well as to educate young Africans with science and technology sense for sustainable and development energy future. Main features and strategy of SSB first stage plan are summarised as follows:

- Solar PV in Sahara desert for future global energy;
- Low cost mass production of solar silicon from desert sands;
- Breeding (doubling) of PV power system every energy payback time (~ 2 years);
- Local PV application: desert greening, new industries, etc…;
- Long distance dc electric transmission by HTSC cable for energy paradigm shift;
- Asia-Africa-Arab sustainable energy cooperation from the beginning of the above basic research to business extension.

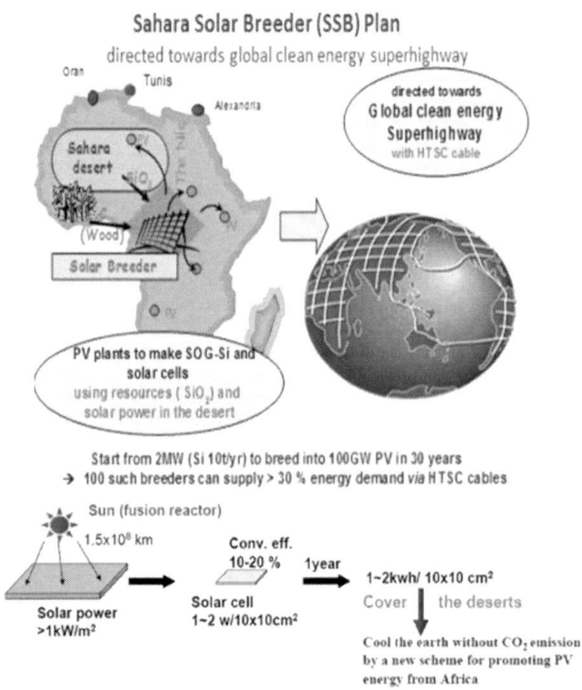

Figure 7. SSB concept, plan, and framework with PV system using the photoelectric effect of turning light energy into electricity.

PV systems have a number of merits and unique advantages over conventional power generating technologies. PV systems can be designed for a variety of applications and operational requirements, and can be used for either centralized or distributed power generation. Energy independence and environmental compatibility are two attractive features of PV systems. The fuel (sunlight) is free and no noise or pollution is created from operating PV systems. In general, PV systems that are well designed and properly installed require minimal maintenance and have long service lifetimes.

The SSB project is therefore serving as the initiative for the super Apollo project (ssb-foundation) to provide the paradigm shift in global energy system. SSB merits and difference from other projects is that:

- High skill jobs are created continuously
- People enjoy to study and work over generations
- Deserts are converted to area for more life
- Happy to manage growth of own economy
- Sustainable distribution of energy to the globe
- Supporting the world sustainability (social responsibility)

SSB Plan and Activities

SSB is a business orientated bottom up collaboration research project between Japan and Algeria. The dynamic interplay of the Triple C (Culture, Community, and Consensus), Triple E (Environment, Energy, Economy) and Triple G (Geology, Geopolitics, Geo-strategy) elements are crucial to the success of the SSB project. The SSB project is not an outmoded technology transfer or academic project. The Japanese must learn many things from the Algerian to achieve a Win-Win strategy scenario between Japan, the Arab and African countries, and it is our destiny to uncover the energy solutions to the world's environment problems with the cooperation and support of the Algerian, Arab and African cultural leaders, economists, finance specialists, politicians, government bureaucrats and the like. SSB project permitted for USTO-MB and Saida University the acquisition of equipment for Training and Research in RE, this offered a unique opportunity of technological advancing in the energy field. The goals of SSB towards sustainable development are founded on four pillars namely:

- Starting from basic research and development;
- Coupling PV with HT_cSC for clean energy generation and energy saving transmission;
- Education and training for science and technology of African people;
- Solving global crisis by international cooperation and policy.

The three basic strategies of the SSERC programmes and indicative targets are based on the following items:

- Innovation processes for solar silicon with a focus on using Sahara sands, the following objectives have been selected:
  - Purification of silica sand by a thermodynamics based process designed in Tokyo and Nihon universities;
  - Reduction of silica sand by carbothermal (Hirosaki University) and plasma (National Institute for Materials Science, NIMS) processes;
  - Technology transfer to USTO.
- Quantitative data collection for installation of PV and HT_cSC systems, the following objectives have been selected:
  - PV system installation at Saida site;
  - Data collection for DC power transmission trough desert environment.
- Education and training on Energy problems and sustainable development, the following objectives have been selected:
  - Installation of WebEl system at USTO and Saida universities;
  - Development of PV materials devices and system;

- PV application such as desalination, green innovation in desert;
- Personal movement between Algeria and Japan.

The SSB knowledge platform is an initiative of the SSB foundation that brings together people, communities, and organisations worldwide to work together on issues relevant to the SSB concept. Moreover, SSB is an integrated community, socio-economic, industrial, agricultural, environmental, and science and technology development solution. In particular, through desalination and adequate irrigation technologies, it will help mainstream marginal desert water resources and lands back into national, regional and international development processes as shown in Figure 8 (BP Statistical Review of World Energy, 2014). Scenarios help us understand the limitations of our 'mental maps' of the world – to think the unthinkable, anticipate the unknowable and utilise both to make better strategic decisions.

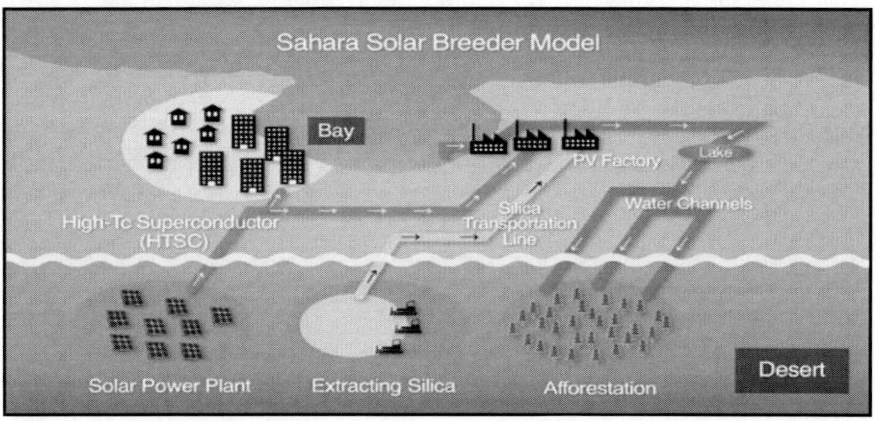

Figure 8. SSB model, key to a sustainable civilization.

Sahara Solar Breeder Components

The most important objectives of SSB's energy policy and its portfolio include five basic strategies that are:

- Basic, applied, practical research and development, in Japan, North Africa, the Middle East, Africa, and other regions of the world;
- Industrial production of silicon from sand;
- Industrial production of cells, modules, panels, and other PV devices,
- Building, operating, networking and monitoring Very Large Scale Photovoltaic Power Stations (VLS-PVPS);
- Environment monitoring and gradual implementation of SSB.

One of the strengths of PV is to be found in its decentralised applications. This is particularly true for supplying isolated consumers in areas of low population density, where the demand consists essentially in satisfying basic energy requirements. Other notable characteristics of PV are:

- Modular design enabling it to be extended according to need;
- The possibility of developing small businesses in areas of low economic development;
- Protection of the environment;
- Limited capital assets, capable of being used flexibly and in a decentralised way, and of being moved about over longer periods of time.

The developing strategy, by the SSB project, has been elaborated to promote the dissemination of RE on sites where they are profitable compared to classical energies and to guide scientific research efforts in order to allow generalisation of RE via mass production. The aims to be achieved consist of the contribution to a conservative policy for hydrocarbons both by increasing the RE share within the international energy balance and by improving the living conditions of isolated communities. The first operation of installation of PV plant in Saida city would allow on one hand the electricity supply and on the other hand to collect information about:

- Equipment behaviour in Saharan environment;
- Matching the systems with the electricity supply;
- Maintenance of organisation and management;
- Technical -economic system optimisation.

Sahara Solar Breeder Plan for Global Energy System Innovation

Material and energy are key components of our life and indispensable for sustainable development of our world. There are two renowned big benefits of Sahara and other deserts from the viewpoint of energy utilization, i.e., vast land and ample sun shine, which attract much interest for harvesting new clean energy from deserts. The SSB project we proposed and started initiative R&D with Algeria is aiming energy harvest from the desert by extracting the extra value from the sand through the cooperative research with Saharan countries. We have been working for proving the feasibility and superiority of SSB-PV to provide an absolute solution of sustainable development for future world (H. Koinuma et al., 2013). The first key step technological task is the development of solar grade Si from desert sand, development based on thermodynamic calculations. Sahara has a plenty of silica as primary material for silicon and of sun shine as solar energy source. Table 1 summarises solar cell materials and power generation capability (evaluated from material needs, efficiency, and natural resources).

Silicon dioxide is the most abundant natural resource on Earth. If we could develop an efficient process for reducing SiO_2 to solar-cell grade silicon (purity level > 99.9999%) and use this to make polycrystalline (poly-silicon) solar cells, solar energy could be converted into electricity in about 15% efficiency. This efficiency appears to be lower than conventional power-generation systems (about 40% efficient), but it is substantially higher in view of the fact that sunlight is fuel free.

Silicon is the second most abundant element in the earth shell, much more than iron, but Si production scale is far lower than Fe, especially of semiconductor grade (SEG-) by 4 to 5 orders of magnitude. The over-specified quality (>10 nines) of SEG-Si had been the main reason of high-cost of Si-solar cell. Despite the recent temporal low price of Si by the overproduction in China, the development of new SOG-Si (~6 nines purity) process with minimum energy consumption is presumed to be the first priority for realising clean sustainable energy future. It must be noted that SOG-Si PV is more promising than not only fashionable i.e., so-called innovative and next generation PV but also conventional power stations (H. Koinuma K et al., 2013). Si renaissance strategy is derived from pioneering but forgotten research in 1980s on SOG-Si from silica sands (H. Koinuma et al., 2015) to make it smarter by the Beyond Siemens process shown in Figures 9 and 10.

Table 1. Solar cell materials and power generation capability

Material	Type	Thickness (μm)	Conversion efficiency (%)	Element	Wp/g	Resource (10^3 tons)	Capability of power generation	
c-Si	W	200	20	Si	0.1	∞	∞(GW)	
	TF	20	15	Si	0.75	∞	∞(GW)	
A-Si	TF	0.7	10	Si(H)	19	∞	∞(GW)	
InP	W	200	25	In	0.33	1.68	0.56	(GW)
	TF	2	20	In	26	1.68	44.5	(GW)
GaAs	W	200	25	Ga(As)	0.49	110	53.9	(GW)
	TF	2	20	Ga(As)	39	110	4,310	(GW)
$CuInSe_2$	TF	2	15	In (Cu)	38	1.63	64	(GW)
				Se	28	83	2,290	(GW)
CdTe	TF	2	15	Cd	27	555	15,100	(GW)
				Te	24	22	526	(GW)
Ge	W	200	15	Ge	0.14	44	0.62	(GW)
	TF	0.5	15	Ge	60	44	250	(GW)

W: Wafer, TF: Thin Film.

Among various PV systems, only Si-PV, with only 4% of desert areas, could cover all the energy needs in the world. The collaborative industry (Taiheiyo cement Co. Ltd in Japan), developed a process for purifying amorphous silica sand in Japan by the original wet process of dissolution in aqueous NaOH solution followed by the neutralization to precipitate SiO_2. This process could not beapplied to desert sand directly due to low solubility of the sand which is composed primarily of crystalline silica. We found this problem could be solved by applying the ball-milling to the sand which had been gravimetrically separated and rinsed with HCl in advance. Thus, we could merge the purification process of desert sand with that of amorphous silica as shown in the following figure.

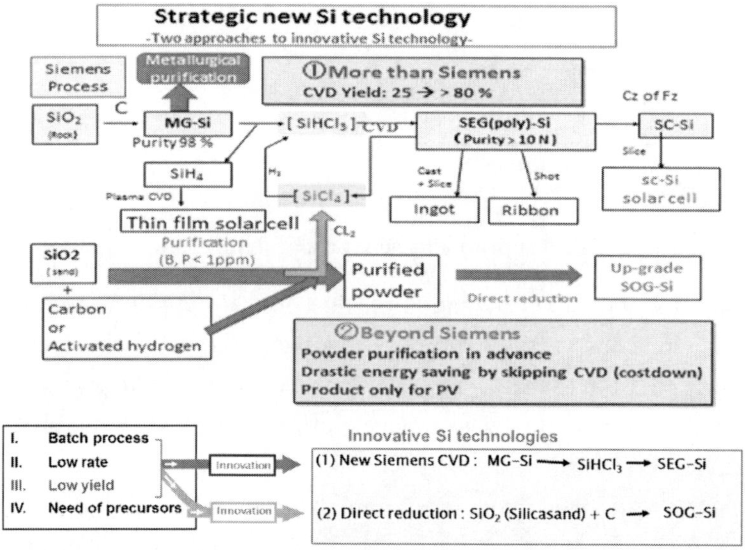

Figure 9. Strategic new Si technology, two approaches to innovative Si technology.

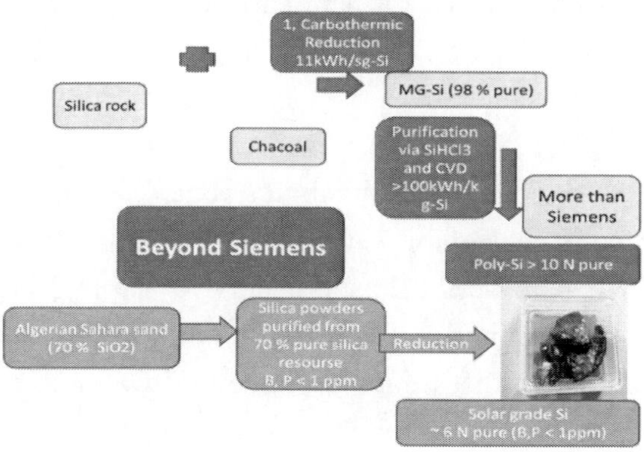

Figure 10. Innovative Si technology.

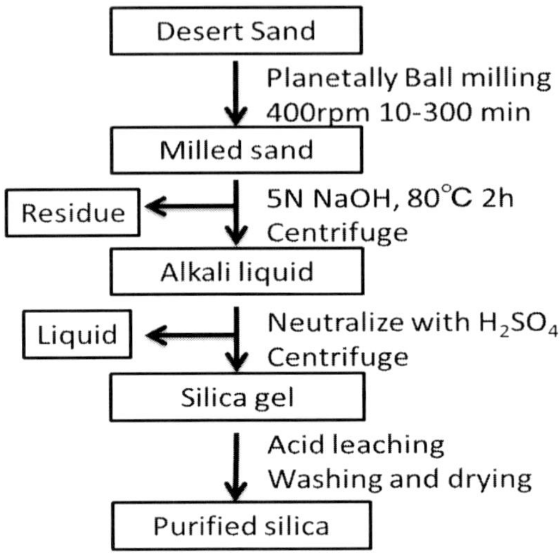

Impurities (ppm)

	Milled	Purified
B	1.8	<0.1
P	52	<0.1

Figure 11. Diagram of sand purification and results by ICP-OES.

Figure 11. Inductively coupled plasma optical emission spectrometry (ICP-OES) analysis verified sub-ppm levels of B and P impurities in the silica purified from Sahara sand at Adrar location in Algeria. Metallic impurities are known to be segregated effectively by the unidirectional solidification process.

In 1980s, NEDO-supported Nippon sheet glass/JFE Co. joint project, which reported that purified silica reduction with carbon powder in an arc furnace using graphite shafts to feed pelleted SiO2 and C to give 6 nine grade SOG-Si. This Si was used to fabricate poly-crystalline Si solar cells. Energy conversion efficiencies higher than 12% was achieved (NEDO, 1987). It must be noted that this efficiency is higher than the efficiencies of conventional thin film solar cells. Hence, we can get Si solar cells with reasonably high efficiency and durability, once we have SOG-Si. By omitting the energy consuming CVD process (~100 kWh/kg-Si) in the conventional Siemens process, drastic energy saving and production cost reduction should be achieved. Recently, a new method for solar-grade silicon attracts attention, that is, at first, natural Silica sand resources (such as desert sands, diatomaceous earth, and so on) are highly purified as silicon dioxides by chemical process under low temperatures and then the carbothermic reduction process with keeping their purity. Since this process can carry out the high reduction process of energy cost only once, it can be suppressed total cost for solar cells (K. Itaka et al., 2015). Recently, the carbothermal reduction of silica has a new solution using innovative technology, so-called microwave, and to be taken to achieve the best in energy consumption performance for the production of silicon, these were supplied by means of variable power of 2.5 KW and magnetron operating at (2.45 ± 10) GHz. The new microwave irradiation process has specific advantages in terms of speed, energy efficiency / performance simplicity (A. Aslani et al., 2013).

Sahara Solar Breeder Foundation, International

The SSB concept is globally orientated. Every country has an RE potential that can be developed and to fulfil its mission, Sahara Solar Breeder Foundation, International (SSBF) will engage in promoting and advancing SSB related initiatives such as:

- SSBA: Academic and research initiatives
- SSBB: Business and public policy initiatives
- SSBC: Cultural and public relations initiatives

SSBA and SSBB will undertake academic, research and business relations initiatives in order to create:

- Academia research which is the key in furthering the RE sector. Multidisciplinary research aspect of SSB is vital to this sector's progress;
- Newer metal industry (Silicon, etc.)
- Cement industry;
- Iron industry;
- Carbon industry;

- PV manufacturing Industry;
- Water industry;
- Local engineering company;
- Local project management company;
- Construction industry to realise above industries, etc.

SSBC will also undertake cultural and public relations initiatives in order to:

- Promote solar energy technology that play a fundamental role in the fast-paced implementation of the SSB concept;
- Promote the socioeconomic benefits associated with developing PV solar energy;
- Increase acceptance of PV electricity and raise awareness of PV PS potential, as well as RE and desert potential by informing, educating and outreaching the general public, the political, business and civil society establishments on the global, international, regional, national, local, corporate and civil society level;
- Organise seminars, conference and round tables as well as other forms of public relations;
- Award prizes to individuals, groups or institutions for undertaking pioneering projects or initiatives advancing the implementation of SSB vision and projects;
- Help improve the image of desert life and regions, and build better awareness of their economic potential and cultural heritage;
- Uphold and defend social and ecological standards and environmental and desert regions development and protection;
- Create a supportive environment for RE investment and development by the SSB;
- Find financing mean of RE projects which is critical in ensuring that projects such as SSB go from conceptualisation to implementation.

SSBB suggests that investment is a key point for diffusion of RE technologies. Specifically, to utilisation of RE economically reasonable, they should be adopted pervasively by supports of the government and contributions of the private sector. Research show that financial measurement that indicates the required investment and other costs of RE utilisation (e.g., maintenance and operation), as well as efficiency of each energy source (performance) are two key criteria for RE promotion (K. Tubouchi et al., 2015). For instance, today the cost of solar PV technologies is being fall quickly as demand is rising with costs declining by 19% with each doubling of global capacity (OECD, 2012). SSB may then acts as a potentially significant new mean of jobs and rural growth in developing countries, and as a means of addressing environmental and energy security concerns. Moreover, SSB is also a science based bottom up approach industrial oriented project as depicted in Figure 12 (A. Boudghene Stambouli, 2015), (*1 = IPCC (Intergovernmental Panel on Climate Change, *2 = COP-FCCC, Conference of the Parties Framework Convention on Climate Change). SSB is therefore a solution to provide social and environmental benefits. To succeed diffusion programs of SSB project development, different strategies such as technological improvements, increased economies of scale, and strong policy support should be contributed in both developed and developing countries.

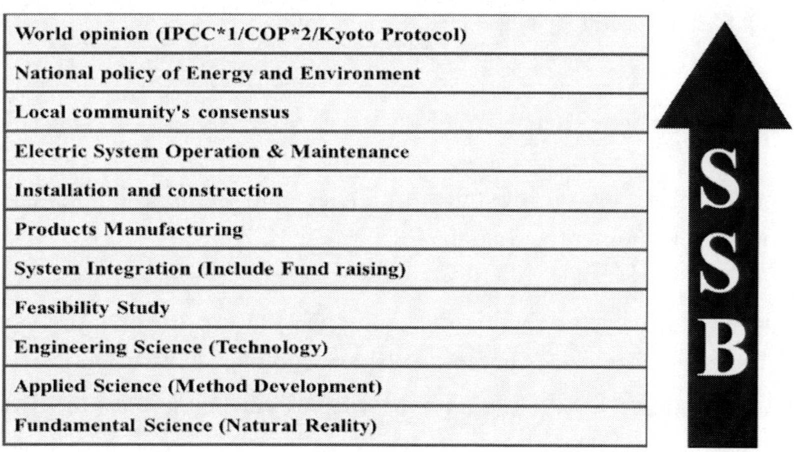

Figure 12. Bottom up approach of SSB project.

SSB Proposal for the Second Stage and for Prospecting the Future

Despite the continuous efforts made during the 05 working years of SSB project to achieve and improve its initial strategy, it is not yet utilised to its maximum potential. Government, non-government agencies and the public will have to take a more proactive step to promote the project. We must strive to increase our efforts in attaining greater efficiency in energy conversion, transmission and utilisation through the second step of SSB project. SSB project future targets starting from Algeria are depicted in Figure 13.

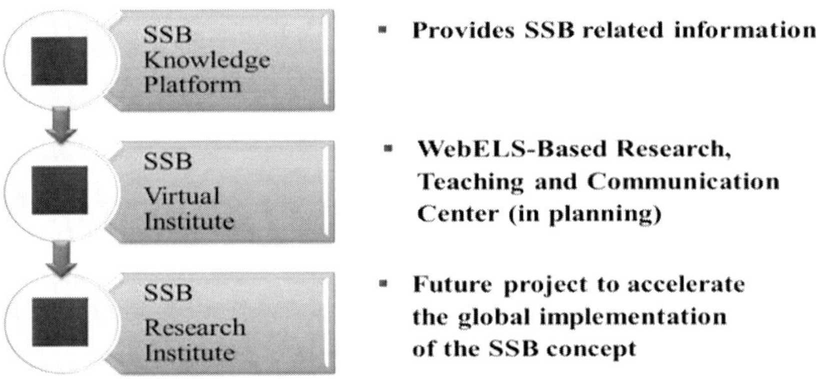

Figure 13. SSB project future targets.

Thoughts for the SSB Research Centre, SSERC

They require us to implement policies and strategies adapted to sustainable development with an SSERC being:

- A think tank for finding the optimum roadmap to solve the world energy problem at the required speed and in a stable way;
- A platform for linking natural sciences and humanities to discuss the development and implementation of renewable energy concepts;

- A forum to exchange ideas and to identify the vital questions or problems concerning the future energy system;
- A host that helps to form international collaborations between research groups or research institutes that can solve the above questions;
- There is an immense need to get things going – and to get them going fast;
- There are many ways to go – we should discuss and find the best options;
- SSB Foundation could be "stand-alone" or in some relation to other academic and research activities of SSB or a "seed" of the SSERC, or even more …

Trends in Sustainable Energy Research by SSB

The main mission for second stage is to ensure the implementation of the SSB Solar Plan and the implementation of the government and SSB policy on RE and PV in particular which requires a significant participation of the academic and research institutions.

The involvement of academic, research and business institutions and their resources is essential to achieve optimal benefits of this second stage.

Highly qualified professionals including designers, engineers and technicians will be required to contribute to this second process as specific factors to bear in mind are shown in Figure 14.

Some key innovative technologies were developed during the first stage, and more should be challenged together with a new concept:

- Support to the administration through the proposition of national and regional plans and incentive measures for RE and energy efficiency and transport development (USTO-MB/Sonelgaz-Sonatrach);

- Identification, evaluation and realisation of the cartography of resources in RE and potentials of energy efficiency and proposal of development zones for solar energy;
- Support to R&D and continuing vocational training on solar energy;
- Mobilisation of the national technical and financial co-operation and participation in companies developing RE activities;
- Provide R&D services to partners, particularly in the fields of environment, energy, and valorisation of Algerian natural resources;
- Implementation of arrangements to develop, coordinate and enhance the efficiency of research in the fields of solar energy;
- Valorisation and dissemination, with Japanese colleagues, of PV solar related research projects results;
- SSB portal will be an entry point for scientists, students, teachers and otherwise interested parties to who want to actively contribute to the development and dissemination of the ideas of the SSB concept. It will give direct links to the relevant information and special interest groups within the SSB foundation.

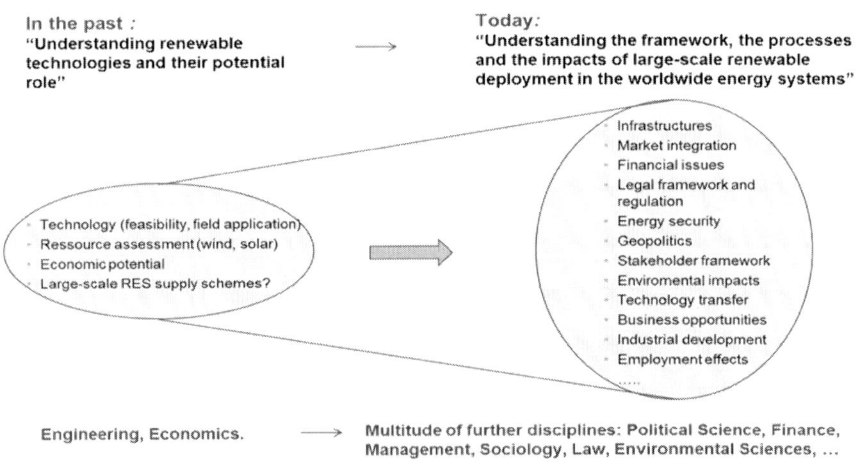

Figure 14. Trends in sustainable energy research by SSB.

USTO-MB–BEYOND SSB FIRST STAGE PROJECT

The new key challenges for the SSB partners groups in Japan and Algeria are putting to work on new raw material, and focussing on the Diatomite.

Diatomite is a well-known natural product; it carries several names such as diatomaceous and kieselguhr. The geological reserves, in Algeria, are estimated at 6.5 million of tones. The diatomite ore is extracted manually underground. It is a light coloured rock consisting primarily of amorphous silica at a rate of 80%, the rest consists of percentages of quartz and impurities of metal oxides, carbonates and salts. The tam of USTO-MB is now investigating the purification of diatomaceous earth to a level of purity 4N, using only NaOH dissolution process followed by washing with sulphuric acid. Applying various protocols on a single purification cycle, by controlling the rate of silicate over pure silica, shows improved purity to a 4N grade.

In addition, concerns such as energy transport have caused researchers from Japan and Algeria and governments to debate about security and diversification in their energy transport policies. As future research, different scenarios such as DC superconducting power transmission line as the most efficient energy transport and a proposal of electric transmission from Algeria to Europe can be defined to compare with energy transport development in Algeria.

CONCLUSION

The importance of access to clean energy cannot be overstated. Research, Development and Demonstration (RD&D) through the SSB project are the foundation for progress and change toward sustainable energy systems throughout the world, protect the global life-supporting systems and reduce the risk of geopolitical conflicts over fossil fuel resources which are quickly-depleting materials. We need more electricity from RE, more investment and initiative to make PV solar energy cost effective and competitive as suggested by SSB concept for future. There are good examples of business initiatives to enhance energy security for communities in energy stressed regions and the lessons of good practice, from experienced countries, such as Japan, need to be more widely disseminated. Cooperation is crucial; problems cannot be solved by one country alone. We need to learn from each other- local, global, regional, cooperation is therefore crucial.

Clean and secure energy =

- Key to a Sustainable Civilisation;
- Key to Human Security;
- Key to a Reliable Energy Supply;
- Key to Climate Stability and Bio-Diversity;
- Key for Global and Inter-Generational Justice;
- Key to more Civilisation and Wealth.

SSB =

- Energy/Climate security with global justice and development of civilisation for whole world;
- Clever global development strategy for solving the energy and climate problems with existing solar grade Silicon production from Sahara sand technology for a world in a sustainable way.

The created SSB system dynamics model, in Algeria, can be implemented in other countries, such as Turkmenistan, and the results can be compared with the current work performed in Japan and Algeria. Further, accurate analysis of each parameters of SSB concept utilisation, strategies for raw material cost reduction along with other factors such as Leading Development programs related to the state policy and the needs of the country in the field of energy conservation and efficiency, RE, energy transport… and combination of energy market, are subjects that are suggested by the authors for future (see diagram of Figure 15) A. (Boudghene Stambouli et al., 2015). The share of installed RE capacity in

Algeria will totalise 40% of total installed electricity generation capacity by 2030. Algeria has now opted for one technology: PV (M. Belarbi et al., 2016).

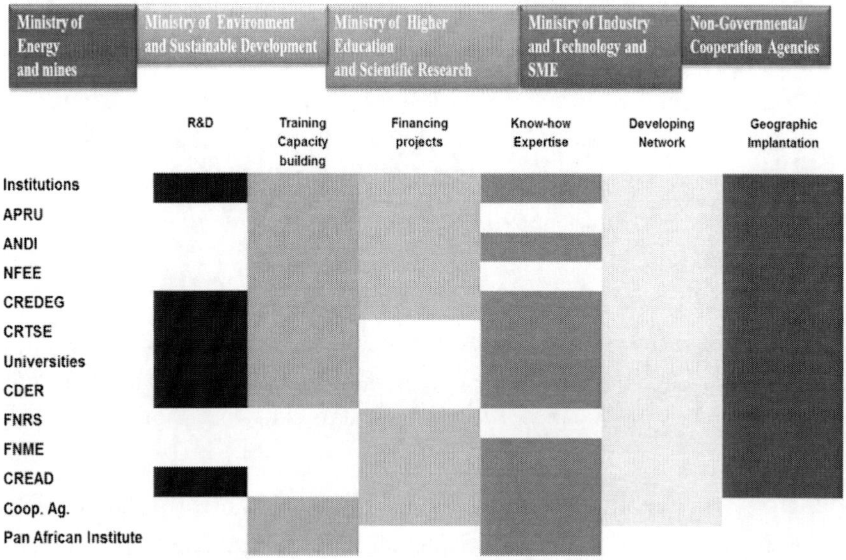

Figure 15. Status /Repartition of Algerian Institutions operating in the RE field.

REFERENCES

Abdallah, M. A., Lei, Z. M., Li, X., Greenwold N., Nakajima, S. T., Jauniaux, E., Rao, C. V., 2004. Human Fetal nongonadal tissues contain human chorionic gonadotropin/luteinizing hormone receptors. *J. Clin. Endocrinol. Metab*. 89, 952-956.

Aslani A., Helo, P. and M. Naaranoja, 2013, Evaluation of renewable energy development in power generation. *Journal of Renewable Sustainable Energy* 5, 063132 (2013).

Belarbi, M., Boudghene Stambouli, A., Belarbi, H. and Haddouche, K., 2016, A new algorithm of parameters estimation of a photovoltaic solar panel. *Turkish Journal of Electrical Engineering and Computer Sciences*, 24 (1): 276-284 (2016).

Boudghene Stambouli, A. and Koinuma, H. 2012, A primary study on a long-term vision and strategy for the realisation and the development of the Sahara Solar Breeder project in Algeria. *Renewable and Sustainable Energy Reviews,* 16 (2012), 591– 598.

Boudghene Stambouli, A. and Koinuma, H. 2015, The Sahara Solar Breeder (SSB) Project Contributes to Global Sustainable Energy Production and Resource Conservation: An Overview. Environmental Sustainability: Role of Green Technologies, DOI 10.1007/978-81-322-2056-5_6, © Springer 2015, ISBN 978-81-322-2055-8, ISBN 978-81-322-2056-5 (eBook). *Springer New Delhi Heidelberg New York Dordrecht London.*

Boudghene Stambouli A., Flazi, S. and Koinuma, H. 2012, Sustainable development by Sahara Solar Breeder plan: Energy from the desert of Algeria, a Green Energy Dream grows in the Sahara. SolarTR-*2 Solar Electricity Conference and Exhibition* November 7-9th, 2012, Antalya, Turkey.

Boudghene Stambouli, A., Flazi, S. and Koinuma, H. 2014, Renewables within the Algerian Energy System: Status, Experiences, Challenges and Opportunities in the Framework of the Sahara Solar Breeder Project. EFEA 2014: *3rd International Symposium on Environment-Friendly Energies and Applications*, 19-21 November 2014, Supmeca, Saint-Ouen, Paris, France.

Boudghene Stambouli, A., Flazi, S. and Koinuma, H. 2015, Powering the planet with solar fuel: SSB project challenges in solar grand plan energy utilisation. *The 5th Asia-Africa Sustainable Energy Forum Jointly with 7th International SSB Workshop*, Tsukuba University, Japan, May 11-13/2015.

Boudghene Stambouli, A., Khiat, Z., Flazi, S., Tanemoto, H., Nakajima, M., Isoda, H., Yokoyama, F., Hannachi, S., Kurokawa, K., Shimizu, M., Koinuma, H., Yassaa, N. 2014, Trends and Challenges of Sustainable Energy and Water Research in North Africa: Sahara Solar Breeder Concerns at the Intersection of Energy/Water. *Renewable & Sustainable Energy Reviews,* Vol. 30C (2014), pp. 912-922.

Boudghene Stambouli, A., and Koinuma, H. 2015, Environmental Sustainability: Role of Green Technologies, DOI 10.1007/978-81-322-2056-5_6, © Springer 2015, ISBN 978-81-322-2055-8 ISBN 978-81-322-2056-5 (eBook) *Springer New Delhi Heidelberg New York Dordrecht London*. The Sahara Solar Breeder (SSB) Project Contributes to Global Sustainable Energy Production and Resource Conservation: An Overview.

BP Statistical Review of World Energy, June 2014. bp.com/ statisticalreview.

ExxonMobil, 2012, the outlook for energy-a view to 2040. *ExxonMobil Corporation, January 2012*. www.exxonmobil.com/energyoutlook/.

Itaka, K. 2015, High-purity Carbothermic Silica Reduction for Solar-Grade Silicon. *The 5th Asia-Africa Sustainable Energy Forum Jointly with7th International Workshop on Sahara Solar Breeder*. May 11th –13th, 2015, University of Tsukuba, Japan.

Kadir, M. Z. A. A. and Rafeeu, Y. A. 2010, Review on factors for maximizing solar fraction under wet climate environment in Malaysia. *Renewable & Sustainable Energy Reviews,* Vol. 14 (2010), pp. 2243–2248.

Koinuma, H. and BoudgheneStambouli, A. 2014, Science and Technology Cooperation between Japan and Algeria Initiated from University of Science and Technology of Oran-Strategy and Current Status of Sahara Solar Breeder (SSB) Plan-. *Book of Sustainable North African Society, Exploring Seeds and Resources for Innovation*. Editors: H. Isoda, M. Neves and A. Kawachi. Published by Nova Sciences Publishers, Inc. New York, 2014.

Koinuma H., and Boudghene Stambouli, A. 2013, "Current status and future plan of SATREPS-SSERC (Sahara Solar Energy Research Center) project." *3rd AASEF (3rd Asia-Africa Sustainable Energy Forum) Jointly with 5th International SSB Workshop*, Hirosaki, Aomori, Japan, May 6-8/2013.

Koinuma H., Tsubouchi, K., Fujioka, H., Morita, K., Stambouli, A., Sumiya, M., Itaka, K., Furuya, Y., Miura, K., Jeong, S. Y. 2011, Feasibility study of Beyond Siemens process for solar silicon from desert sands. *Material Research Society* MRS 2011 Fall Meeting Symposium Africa, Zimbabwe, 06-10/12/ 2011.

Koinuma, H. 2015, Progress and potential of SSB project for sustainable energy future. *The 5th Asia-Africa Sustainable Energy Forum Jointly with7th International Workshop on Sahara Solar Breeder*. May 11th – 13th, 2015, University of Tsukuba, Japan.

Koinuma, H., Kanazawa, I., Karaki, H. and Kitazawa, K. 2009, *Science Council of Japan proposal at G8+5 academies meeting*. March 26-27, 2009. http://www.scl/go.jp/ja/event/pdf/79-s-3-1.pdf.

Koinuma, H., Tsubouchi, K., Itaka, K., Stambouli, A. 2013, "Sahara Solar Breeder project," *McGraw-Hill 2013 Yearbook of Science and Technology*, pp. 325-330, 2013.

Kurokawa, K., Keiichi, K., Masakazu, I., V. Peter Van Der, F., 2009, David, editors. *Energy from the desert- III*: IEA-PVPS task-8 report. UK: James & James Ltd.; 2009.

NEDO report, http://www.nedo.go.jp/library/seika/shosai_201003/20100000000220.html, 1987.

OECD, 2012, Linking Renewable Energy to Rural Development, *OECD Green Growth Studies*, OECD Publishing.http://dx.doi.org/10.1787/9789264180444-en.

Ssb-foundation, http://www.ssb-foundation.com.

Tubouchi, K., Toyokura, N., Ishihara, S., Watanabe, T., Hayashi, M., Nagata, K., Masuda, K., Miura, K., Shimizu, M., Koinuma, H. 2015, Direct reduction of high purity silica by microwave heating. *The 5th Asia-Africa Sustainable Energy Forum Jointly with7th International Workshop on Sahara Solar Breeder*. May 11th –13th, 2015, Tsukuba University, Japan.

Yamaguchi S., Iiyoshi, A. and Koinuma, H. 2015, DC Superconducting Power Transmission Line as the most Efficient Energy Transport & A Proposal of Electric Transmission from Algeria to Europe. *The 5th Asia-Africa Sustainable Energy Forum Jointly with7th International Workshop on Sahara Solar Breeder*. May 11th– 13th, 2015, University of Tsukuba, Japan.

Zerdali M., Benhadjaj, S. M., Medeghri, C., Chahtou, I., Hamzaoui, S. Toward of Refined purification of Algerian diatom for solar grade silicon. *The 5th Asia-Africa Sustainable Energy Forum Jointly with7th International Workshop on Sahara Solar Breeder*. May 11th– 13th, 2015, University of Tsukuba, Japan.

BIOGRAPHICAL SKETCH

Amine Boudghene Stambouli

Affiliation: University of Sciences and Technology of Oran (USTO-MB)

Education: Engineer, Master, PhD

Business Address: BP 1505, EL M'Naouer, Oran (31000), Algeria. Tel & Fax: 00 213 41 627163

Research and Professional Experience: Started working on environmental friendly production of energy and sustainable development related issues. In 2001, he moved to the University of Roma Tor-Vergata to take up an academic appointment as a University Researcher at the Chemistry and Material Science Laboratory with Pr. Enrico traversa who was Co-responsible of the research team « Photovoltaics and fuel cells » created by Amine Boudghene Stambouli. In 2002, he was promoted to full Professor of Optoelectronics and Material Science for Environment and Energy Applications at the Department of Electronics of USTO-MB. Amine's current research projects include Photovoltaics, Fuel cells, Hybrid Systems, and Environment Impacts. Amine is a United Nations consultant (Index 382958), member of many scientific and industrial organisations and director of several doctoral courses and thesis.

Professional Appointments: Served as the head of the electronics department, vice rector of the university for almost two years and later the president of the scientific council of the Electrical and Electronics Engineering Faculty of USTO. In March 2013, he was appointed as the Director of the Electrical Engineering Laboratory of Oran. Amine is also member of the scientific council of Polytechnic University of Oran and the CRTSE in Algiers. He has been chairing several international conferences in the field of Electrical Engineering and was chairman at numerous sessions of international conferences. His studies are documented by 4 books and over 250 papers mainly published on International Journals and on Proceedings of International Conferences. He was awarded the prize of the best publication of the year 2009, delivered by CDER and the Ministry of Higher Education and Scientific Research of Algeria. He is a reviewer and an Editorial Board Member of several International Journals. He is actively collaborating with research group worldwide (Algeria, Italy, Japan, France, USA, Germany, Turkey, England, Saudi Arabia, Jordan, Morocco, Tunisia, Spain, Oman and Syria). He is Manager (Algerian side) of the Sahara Solar Breeder (SSB) project along with Pr. Koinuma (Japan side) and founder of the Sahara Solar Breeder Foundation (www.ssb-foundation.com).

Honors: awarded the prize of the best publication of the year 2009, delivered by CDER and the Ministry of Higher Education and Scientific Research of Algeria.

Publications from the Last 3 Years:

1. *Renewable & Sustainable Energy Reviews (Elsevier)* Vol. 30C (2014), pp. 912-922. Trends and Challenges of Sustainable Energy and Water Research in North Africa: Sahara Solar Breeder Concerns at the Intersection of Energy/Water.
2. *The Arabian Journal for Science and Engineering (AJSE).* June 2014, Volume 39, Issue 6, pp 4911-4921. Multiuser Detection for DS-CDMA Systems using Honey Bees Mating Optimization Algorithm.
3. *Book of Sustainable North African Society, Exploring Seeds and Resources for Innovation.* Editors: H. Isoda, M. Neves and A.

Kawachi. Published by Nova Sciences Publishers, Inc. New York (2015). Science and Technology Cooperation between Japan and Algeria Initiated from University of Science and Technology of Oran-Strategy and Current Status of Sahara Solar Breeder (SSB) Plan-.

4. *Sustainable Practices: Concepts, Methodologies, Tools, and Applications, IGI-global Book* 2014. Study of Hybrid Sustainable Energy System Based on PEM Fuel Cells and Photovoltaic-Module Power Generator. DOI: 10.4018/978-1-4666-4852-4.ch076.http://www.igi-global.com/chapter/study-of-hybrid-sustainable-energy-system-based-on-pem-fuel-cells-and-photovoltaic-module-power-generator/95000
5. Scientia Iranica Transactions D: Computer Science & Engineering and Electrical Engineering D (2014) 21(6), 2273-2279, ISSN: 1026-3098, e-ISSN: 2345-3605. *Modelling polycristallin photovoltaic cells using design of experiments*. http://www.scientiairanica.com/en/ViewIssueArticles?IssueId=40.
6. *Renewable and Sustainable Energy Reviews* 41 (2015) 261–276. A review on the water and energy sectors in Algeria: Current forecasts, scenario and sustainability issues.
7. *Environmental Sustainability: Role of Green Technologies*, DOI 10.1007/978-81-322-2056-5_6, © Springer 2015, ISBN 978-81-322-2055-8 ISBN 978-81-322-2056-5 (eBook) Springer New Delhi Heidelberg New York Dordrecht London. The Sahara Solar Breeder (SSB) Project Contributes to Global Sustainable Energy Production and Resource Conservation: An Overview.
8. *International Review of Aerospace Engineering* 8(4):141, August 2015. Design and Optimization of Microsatellite Power System.
9. *International Journal of Engineering Innovation & Research,* Volume 4, Issue 3, ISSN: 2277 – 5668 (2015). Energy Crisis, Climate Changes, Energy Alternative and Technological Bias of Solar Photovoltaic

10. *Renewable and Sustainable Energy Reviews* 56 (2016) 887–892, Optimization of distribution networks supplied by photovoltaic sources, using a new optimal loads distribution center method.
11. *Turkish Journal of Electrical Engineering and Computer Sciences*, 24 (1): 276-284 (2016). A new algorithm of parameters estimation of a photovoltaic solar panel.
12. *Telkomnika Indonesian Journal Of Electrical Engineering*, Vol. 16, No. 3, December 2015, pp. 488 ~ 494. DOI: 10.11591/telkomnika.v16i3.9421. Prototyping a Dedicated Photovoltaic System Datalogger.
13. *Renewable Energy* 97 (2016) 169-176. Optimization of the investment cost of solar based grid.
14. *Handbook of Research on Modern Optimization Algorithms and Applications in Engineering and Economics*, pp. 581-600 (March, 2016). ISBN13: 9781466696440|ISBN10: 1466696443|EISBN13: 9781466696457|DOI: 10.4018/978-1-4666-9644-0. Parameter Optimization of Photovoltaic Solar Cell and Panel Using Genetic Algorithms Strategy. DOI: 10.4018/978-1-4666-9644-0.ch022. http://www.igi-global.com/chapter/parameter-optimization-of-photovoltaic-solar-cell-and-panel-using-genetic-algorithms-strategy/147530.
15. *Renewable and Sustainable Energy Reviews* 65 (2016) 319–331, A review of the water-energy nexus.
16. *Renewable and Alternative Energy: Concepts, Methodologies, Tools, and Applications, Information Resources Management Association* (USA) (3 volumes), Copyright: © 2017. ISBN13: 9781522516712|ISBN10: 1522516719|EISBN13: 9781522516729|DOI: 10.4018/978-1-5225-1671-2. Book chapter entitled: Modeling and Simulation of a Stand-Alone Hydrogen Photovoltaic Fuel Cell Hybrid System, DOI: 10.4018/978-1-5225-1671-2.ch016, Pages: 545-580.
17. *Renewable and Sustainable Energy Reviews* 73C (2017) pp. 26-43, Prospective Analysis for a long-term Optimal Energy Mix

Planning in Algeria: Towards High Electricity Generation Security in 2062. http://dx.doi.org/10.1016/j.rser.2017.01.023.

18. *Renewable Energy* 107 (2017) pp. 53-60. Performance assessment of five different photovoltaic module technologies under outdoor conditions in Algeria. http://dx.doi.org/10.1016/j.renene.2017.01.057.
19. *International Journal of Engineering Sciences & Research Technology* 6 (2): February, 2017 (ISSN: 2277-9655). Management Strategy of Hybrid Wind/ Solar System. DOI: 10.5281/zenodo.291869. http: // www.ijesrt.com.
20. *Computers and Mathematics with Applications* 74 (2017). PDE-based efficient method for colour image restoration. https://doi.org/10.1016/j.camwa.2017.05.011.

In: North Africa
Editor: Stephen Fuller

ISBN: 978-1-53612-983-0

Chapter 3

DIGITAL POLITICS IN NORTH AFRICA: POSSIBILITIES TO REVERSE SUBVERSION

Ibrahim Saleh, PhD
Nile University, Cairo, Egypt

"The world is not sliding, but galloping into a new transnational dystopia. The Internet, our greatest tool of emancipation, has been transformed into the most dangerous facilitator of totalitarianism we have ever seen. The Internet is a threat to human civilization."

Cypherpunks: Freedom and the Future of the Internet, Julian Assang

ABSTRACT

North Africa is marked with an unmatched phenomena of increase in IT (Adeiza, 2013), though IT users are not homogenous groups, but rather heterogeneous communities with diverse interests and motivations. As explained by the Swiss Agency for Development and Cooperation (SDC) that technology itself cannot solely make a difference in the practice of the sociopolitical and economic progress, but rather the other way around where indicators of development behind the technology drives the change. IT is thus an instrument, not a goal (Sarrazin, 2011).

The synergy between politics and IT is exemplified through many hybrid formations and phases since its initial preparation phase, where activists used digital media over several years to build social networks and identified collective identities. The second launching phase is marked with unprecedented ignition moments, where movement leaders activated these digital networks toward offline protests. And thirdly, profound channeling and programming of information warfare are in place to compete for control, as well as enhance international pressure, economic sanctions, and lobbying efforts (MacKinnon, 2012).

But there are a number of concerns and limitations to what we know, expect and do with IT and digital democratization which coincides with the words of Schraeder (2003), when he described Africa's diversities and complexities as "mosaic in transformation." Such complex and inconclusive crossing roads are further interoperated to serve commercial and political interests in what is called the "Rashmon Effects" (Davenport, 2010).

This makes it rational to mention that the level of legitimacy in using power and establishing identity is the main prevailing trend in Africa, which exploits the social function of media as a driving force for national ideology (Movius, 2010). Social power in many African societies construct meanings in the human mind through communication that utilizes internet and other horizontal digital communication networks (Castelle, 2011).

This research puts forward a key question: "Will the new technologies that have sustained globalization reinforce or undermine democracy?" The aim is to offer a more nuanced approach of deductive forms of investigation about the potential role and impact of digital politics. This difficult path though must be considered by cross examining how politics, digital infrastructure adoption function and affect could contribute to democratization within the African context (Hussain, 2014).

INTRODUCTION

North Africa is marked with an unmatched phenomena of increase in IT (Adeiza, 2013), though IT users are not homogenous groups, but rather heterogeneous communities with diverse interests and motivations. As explained by the *Swiss Agency for Development and Cooperation (SDC)* that technology itself cannot solely make a difference in the practice of the sociopolitical and economic progress, but rather the other way around where indicators of development behind the technology drives the change. IT is thus an instrument, not a goal (Sarrazin, 2011).

The synergy between politics and IT is exemplified through many hybrid formations and phases since its initial preparation phase, where activists used digital media over several years to build social networks and identified collective identities. The second launching phase is marked with unprecedented ignition moments, where movement leaders activated these digital networks toward offline protests. And thirdly, profound channeling and programming of information warfare are in place to compete for control, as well as enhance international pressure, economic sanctions, and lobbying efforts (MacKinnon, 2012).

But there are a number of concerns and limitations to what we know, expect and do with IT and digital democratization which coincides with the words of Schraeder (2003), when he described Africa's diversities and complexities as “mosaic in transformation.” Such complex and inconclusive crossing roads are further interoperated to serve commercial and political interests in what is called the “Rashmon Effects[1]” (Davenport, 2010).

This makes it rational to mention that the level of legitimacy in using power and establishing identity is the main prevailing trend in Africa, which exploits the social function of media as a driving force for national ideology (Movius, 2010). Social power in many African societies construct meanings in the human mind through communication that utilizes internet and other horizontal digital communication networks (Castelle, 2011). Such perfect coupling is motivated by the history of colonization, poverty and socio-political and economic forces to explain, and justify, the role of media as instruments of political power (Movius, 2010).

The social volatility and political upheavals in the African nations reiterates an acute state of the “canons” of intellectual, social, political and economic life. This makes any expectations of IT, democratization and liberation hampered by hegemony, power relations and the primordial identities that are still dominating the African societies (Soyinka-Airewele, 2010).

---

[1] The namesake comes from an Akira Kurosowa film of the same name in which the same story is told from five different perspectives, with very different interpretations.

In addition, inequalities in Africa is so prevalent that involves a complex process of inclusion/exclusion, which is also related to xenophobia, hate, religious prejudice, class, ethnic and racial discrimination. Each of these elements had its strong effects between the dominant discourses that achieve authoritative status and subordinate discourses that are marginalized, or even silenced ('extreme'), which is almost the case in all African nations.

In such perplexity and fluidity, African scholarship is entrapped in "canonical monologue," which only recycles models, theories and logic that correspond to sanctification of Eurocentric racial, economic and gendered assumptions that are caught within the crippling structures (Osundare, 2002). Sadly, many African scholars still presume similar outcomes of the "information superhighway" on the progress and empowerment of global north to be replicated for the "underdeveloped" south (Mabweazara, 2010).

It is not surprising then to find political authorities in Africa use (directly and indirectly) censorship protocols and surveillance softwares to monitor and control citizens' online experiences. It is thus insufficient to just learn to code languages, practice self-censorship, and evade phone conversations altogether to circumvent surveillance (Herrera, 2015). For example, Hossni Mubarak's regime in Egypt coerced the UK-based Vodafone Corporation to turn off cellular networks on its behalf, for it had not designed centralized "kill switch" strategies during the Egyptian uprising (US Department of State, 2012).

Cultural crisis of stagnations, bitterness of the past and colonization, as well as the widespread hate with all the related naming and shaming techniques have only delayed the road to democracy and in some cases have made it even an elusive dream!

Scholars and practitioners are always divided between technology optimism and pessimism. On one hand, many perceive the positive potentials of digital media in the development of new ways of mobilization, participatory democracy and civic engagement (Carty, 2010). On the other hand, others are still skeptical about its utopian and automatic implications and urge the academy to be more cautious and follow critical stance towards the democratic possibilities of Web 2.0 tools (Roberts, 2009). This argument is further supported based on the fact that corporate media maintained its connections through an elitist audiences, while the ordinary citizens are not 'empowered' by the new digital tools (Hindman, 2009).

This research puts forward a key question: "Will the new technologies that have sustained globalization reinforce or undermine democracy?" The aim is to offer a more nuanced approach of deductive forms of investigation about the potential role and impact of digital politics. This difficult path though must be considered by cross examining how politics, digital infrastructure adoption function and affect could contribute to democratization within the African context (Hussain, 2014).

CONCEPTUALIZING DIGITAL POLITICS IN AFRICA

In order to conceptualize IT and digital politics, I would recall Averroes' concept of "Virtual, or potential Intellect" (Mesbahi, 2007) that considers the impact of IT to the "Enlightenment Reason" in the way that both of them encourage progress, innovation and create its public domain. However, it also brings along contrasting implications of new and more important gap between social classes, individuals and nations within Africa and beyond.

This section highlights and deals with the current problem of lack of locality and authenticity of theorization of digital democracy in Africa; lack of underscoring the linkages between collective action repertoire, new communication technologies, and the politics of 'recognition' and 'distribution' in the region. Such culturally entrenched resistance have resulted from the nature of the primordial societies in Africa that closely link IT to falsification and oppression of political beliefs and freedom, and this of course made people turn away from these technologies.

In addition, there is still insufficient studies of IT directives, and whether these directives could help Africa make frontiers porous to erode the parochial limits that hinder profound economic development, or whether IT undermine democracy, by imperiling the national sovereignty that has been democracy's sanctuary (Barber, 2001). Such multifaceted hurdles relate to the difficulty of acquiring the new technological revolution tools (hard and soft), or the sophisticated ability to use them, as well as the consequence of illiteracy and poverty on such use (moral and material).

Against this backdrop, it is important to identify and discuss some of key operational variables such as the context meaning of democracy. It is thus crucial to specify if we mean "pure democracy" or we can relate to the variations of what we think of as democracy?

On can list here many competing theories of direct democracy and indirect democracy, representative democracy and populist democracy, and plebiscitary democracy and strong democracy (Barber, 2001). However, another difficult question comes in relates to the nature of democracy that we are referring to in the light of assessment of technology's impact on politics?

My research paper focuses on the 'politically motivated movement relying on the internet' that utilizes either IT-enhanced or IT-based (Vegh, 2003). The research aims to map the impact of IT as an alternative news source to counter the control of information channels opposed to the interests of the activists in what is usually described as 'mediated mobilization' (Lievrouw, 2011).

The research rationale coincides with what John Nichols and Robert McChesney stated in *The Nation* (2009), when they argued that technology is no doubt has harmed freedom of expression that requires another kind of intervention instead of limiting ourselves to identify the symptoms (Owen, 2015). So 'digital politics' is a term used int he research to emphasize the digitization of traditional politics, as well as the rise of new forms of political life originating in the digital world, such as *Wikileaks* or the Anonymous movement (Postill, 2012).

'Chains of equivalence' is also relate to the hypothesis that this research makes considering the particular elements that represent a signifier that is strongly influenced by the current disparities in Africa, which stipulates 'counter-hegemonic front' of effective opposition of dominant discourses (Dahlberg,2007).

Besides, it is important to relate to the concept of ‘public sphere,’ which was associated with Jürgen Habermas, as a ‘rhetorical token’ (Benson 2005) as a normative notion that guides research away from what is, and towards what ought to be. As pointed out by (Fraser, 1997, 154) 'hegemony points to the intersection of power, inequality and discourse' that explains the socially contingent systems of meaning (Laclau & Mouffe, 2001).

Following the circle writing style, one retraces the significance of “virtual reason” that empowers citizens to participate in and respond to a public discourse without being entrapped or blocked, by colonizing mode of communication (Habermas, 2006). But there is a necessity to follow a plurality approach that appreciates different socio-technical values and mechanisms that weigh the pros and cons of the low entry threshold and ease of use of Web 2.0 tools (Chadwick, 2008).

It is thus pertinent to make logical correlations between IT and democracy, through seven variables: interactivity, temporal structure, social cues, storage, replicability, reach, and mobility (Baym, 2010). Yet we need to weigh the kind of qualities and activities in particular country contexts that are very diverse and topic specific, by understanding the media and political governance that influence these qualities and activities (Morozov, 2012).

This section revealed the state of the theoretical isolation of Africa for multiple overlapping reasons, which only created deliberate enclaves and ghettos of blind spots and provided alienated approaches. The academy has still failed to draw on existing theoretical paradigms and limit themselves mainly to descriptive analysis of collective action is characterized (Wiktorowicz (2004).

IT, Digital Democracy and North Africa

Investments of IT in the African nations expand annually and even overtaking the rest of the world. But there is a lack of context and domestication that contributed further to the North African states' vulnerabilities and understudies. There is also lack of technical and legislative capabilities, which leaves the region without any concrete legislation or framework (El-Guindy, 2014).

The intense situation in the North Africa is compound with the delinquencies in the structural and functional aspects of cyber legislation in between Arab countries and developed countries. Despite most countries acknowledge of the need of cyber laws for promoting the Information Society; nonetheless these same countries lack a full and homogenous package of these laws (Idlebi, 2015).

There are undeniable attempts in the Arab countries in this direction between 2010 and 2014; law enforcement remains a true iceberg in any progress.

In 2010, the *League of Arab States* signed a Convention on Combating Information Technology. In 2011, the *Economic Community of West African States (ECOWAS)* displayed Directive 1/08/11 on Fighting Cyber Crime within ECOWAS. Since 2012, the *African Union* has developed a Draft African Union Convention on the Establishment of a Legal Framework conducive to cyber security in Africa (UNODC 2013, 64). However, there are no specific national laws, or regulators governing data protection and privacy comparable to those found in jurisdictions in the European Union.

The IT revolution in North Africa and in most of the African nations collided with the local traditional cultures and values. And in many situations, there is a lack of cautiousness and vigilance to the leveling and anarchy that IT brought with! This makes the way to move forward is only, by reconsidering all our educational and cultural systems away from the polarized wisdom of euro-centric standardization, or fear from eradicating cultural diversity as a starting block in overcoming such artificial contradictions.

But any potential of 'pitched battles' in North Africa must be considered and processed through a case specific in deterring the temporal and contextual factors to identify the influence of power relations and socio-historic factors in influencing online activism (Franklin, 2004).

Political upheavals in the Middle East and North Africa (MENA), mistakenly called 'Arab spring,' are strictly socio-political and economic reconfiguration of identity wars that limited the idea of Foucault (1986) described as 'heterotopias,[2]' or the establishment of 'safe spaces' that allow for the flourishing of new, potentially dangerous identities that challenge the existing order.

Most literature confirmed the close connections between online and offline activism, and the matching increase of control of authoritarian structures on new media, which explains the correlation between proliferation of IT; online suppression and culture of subversion. There is an undoable evidence of cumulation of interaction between internet activists and more established opposition movements in the lead-up to the protests (Eaton, 2013).

The reasons behind such limitation is due to the cause and motivation in this particular space, in the case of North Africa and all Africa in that regard, through a systematic process of rebelling against unemployment, tyranny and corruption; political, social and cultural conservatism and stagnation, as well as older generations' and elites' preference for the status quo and compromises (Shenker et al., 2011).

Disenfranchised youths have searched for an alternative method of participation in the public and political spheres (Salmon, Fernandez, & Post, 2010). Though, all of the region is marked by pervasive gender inequality that impacts the publics' lives and how they seek to shape society (WLP, 2013). Insurgent politics emerge and dominate the scene and practices its multiplier effects on the social stability.

Such bleak reality erodes any argument that Twitter and Facebook have instigated the revolutions of MENA in 2011, which equally implied a blinkered view on the ground and disregarded many crucial intervening factors.

Caveats aside, it is imperative to look into different platforms and technologies that were used in different ways, though there is no line of

[2] Heterotopias are spaces of crisis, deviance, abnormality and transformation (Kern, 2008: 105) embody a threat to the production of safe space if they are unregulated and uncontrolled.

demarcation between the use, the goals and the functionality when considering digital politics in North Africa.

Having said so, there is a profound error in academic literature when it comes to North Africa in its reliance on euro-centric and non-local insights, research, context, and the overzealous praise of the role of IT has been vocalized by Harvard professor Tarak Barkawi (2011).

> "To listen to the hype about social networking websites and the Egyptian revolution, one would think it was Silicon Valley and not the Egyptian people who overthrew Mubarak."

The research argues that the ideas of 'orientalism' and social distance[3] caused further misconception of interpretations, as well as inadequate understanding of the role of IT played in political activism in North Africa. This might explain how the non-African academics misread the data and IT indicators and were dawn in stereotypical lens as an epithet that connotes 'passivity, unruliness, or propensity to easy manipulation' (Eickelman and Anderson, 2003). Many scholars were not even attentive to the peculiarity of having more mobile phone users than fixed-line subscribers (Goldstuck 2010).

Such marginal dissent in cyberspace mirrors similar processes in physical space to render public spaces in ahistorical, safe and generic that mixed digital activism with jihadist Muslims, where the first group struggle for freedom and seek democracy, while the later destruct churches and demolish historical heritage in Syria and Iraq. Such opaque and confused vision limited the ability of 'resistance identities' to flourish through the evisceration of both symbols and places (Castells, 2004).

"Rashmon Effects[4]" plays a key role in this media hostility and misreading of IT impact on publics, especially youth in North Africa,

---

[3] Social distance determines the conditions under which the third person effect (Davison, 1983) operates. The concept aims to assess the effects accentuated between groups who are remote from the each other. The concept emphasizes the overestimation of media effects is greater when people imagine the responses of others whom they do not know than those of familiar individuals. The third person effect operates in the absence of perceived persuasive intent, but is accentuated when bias is perceived by viewers (Gibbon & Durkin, 2006).

[4] The namesake comes from an Akira Kurosowa film of the same name in which the same story is told from five different perspectives, with very different interpretations

which describes how the political agenda dictates its terms on the reporting source, and be cognizant of variations in the media reports, when constructing data sets from conflicting sources.

These impediments and other related challenges reflect how the Patron states create deliberate enclaves[5] and use all their powers to block access to information related to politics, social issues, and human rights. Politics is being strangled in North Africa through a systematic institutional process of arbitrariness and opacity surrounding decisions to restrict particular content with even expelling the reasons behind such censorship. This mechanism implements state of the art technical attacks to disrupt activists' online networks, eavesdrop on their communications, and cripple their websites.

In many incidents political authorities resort to cyber-attacks when their power is threatened or at stake, by exploiting infrastructural controls to protect their rule, even if it causes massive disruptions to economic activity and personal communications. In other cases, governments employed cruder tactics to delete and manipulate politically or socially relevant information. Such methods are often ingenious in their simplicity, in that their effects are more difficult to track and counteract than ordinary blocking.

THE VALIDITY OF DIGITAL POLITICS IN NORTH AFRICA

Communication power is propounded in North Africa, in particular Egypt and Tunisia, which enabled internet activists to establish networks of resistance (Eaton, 2013).

In this section, the research explains the dystopian potential of technology, by looking into how social media have assisted (or not) the

[5] Deliberative enclaves: Refers to the failure to participate on equal terms as a result of exclusion and disempowered speakers and discourses. However, some argue that democratic equality is best achieved when disempowered groups deliberate in their own enclaves (interest groups, parties, and movements) before entering the broader public sphere. They support this claim by presenting a diversity of viewpoints rather than falling into groupthink and polarization, and can persuade external stakeholders of the legitimacy of the group's deliberations (Karpowitz, Raphael and Hammond, 2009).

digital politics in North Africa. The region is a true case of "youth bulge," with young people far outnumbering older adults (65 percent) less than fourth-years old with (30 percent) under the age of thirty with high educational levels. This blend of education, desperation and unemployment make them the perfect breed to voice out and reflect their frustrations and dissatisfaction with daily endless problems such as unemployment, tough economic conditions and government corruption (Tung 2011).

There are more than (233) universities, (10 million) students, and more than (1000) centers of scientific research in North Africa without being able to make sense of IT and transfer the scientific capacities into efficient capabilities in the world of ICT (ECOSOC Annual Ministerial Review, 2013). Based on a recent ILO report, over (60 percent) of educated youth in the region are in neither school nor work!

In contrast, youth movements are neither represented by structured NGOs or institutions, nor do they form homogeneous groups. Instead, youth movements are linked with a collective agreement on 'a set of opinions and beliefs in a population representing preferences for changing some elements of the social structure or reward distribution, or both' (Zald and McCarthy, 1987).

Youth movements are characterized by a high level of cross-membership and fluidity of structures with a mixed membership from various existing NGOs and political parties that do not adhere to the same ideology or class. However, youth movements such as the *6 April movement* in Egypt and *20 February movemen*t in Morocco play leading roles in digital politics, though they pertain very loose structures.

Young movements in North Africa have an identity crisis and are redefining their loyalties and goals. Youth sections inside political groups, such as the *Youth Muslim Brotherhood* and *Salfi Costa* in Egypt, as well as the *Socialist Union of Popular Forces party* in Morocco, defied their senior leaders and shared common beliefs and opinions rooted in demography, world vision and everyday life experience.

Such youth bulge have utilized the exponential growth in telephone access is unmatched: less than (3 percent) of the population had access to a telephone in 2001, but by 2010 the number of mobile subscribers had grown to approximately 500 million (Rao, 2011). By January 2004, there were (5-8 million) email users in Africa, and around (52 million) mobile phone subscribers, (450 million) SMS messages sent in December 2002, compared to (350 million) for December 2001 (Rao, 2011).

Digital activists recreated the network public spheres with the aim of informing, mobilizing, stipulating robust communities, expanding transparency, and taking control and shift powers from the patron states to the individuals (Ghannam 2011). As identified by Ebenezer Obadare (2004): "while mobile telephony has no doubt come to be seen as a veritable instrument of political struggle, its potential effectiveness is bound to be determined by the way in which it is used".

North Africa has one of the highest rates of technology adoption amongst all developing nations (Howard, 2011), which offers golden opportunities to unite democratic efforts, boost political activism and stipulate notions of freedom and democratization both within specific respective countries and with others nations in the region and beyond (El-Nawawy & Khamis, 2012).

According to *The Global Evolution of Digital Commerce and MENA eCommerce* (2013), internet users in the MENA region, mainly North Africa, reached (77 million) in December 2011according to the Internet World Stats). If we want to give specific country cases, we would find Egypt to be ranked No.2 with over (21.6 million) Internet users; Morocco ranked No.3 with over (15.6 million) users; Algeria ranked No.7 with over (4.7 million) Internet users; Tunisia ranked No.10 with over (3.8 million) users. In general, the total number of internet users in North Africa has reached (46 million) in December 2011.

However, the penetration of the electronic media (new media) remains relatively low, though it is rapidly on the rise. About (1.3million) use Twitter, while just over (40millions) are on Facebook. Facebook penetration stands at (27,711,504), though it ranges across the region from nearly (40 percent) in the Emirates to (12 percent) in Egypt and barely (5 percent) in Iraq (The Economist, 2012).

According to *Arab Social Media Report*, Egypt dominates the scene with its (6,586,260) active Facebook users, which puts Egypt as the highest growth in users in the Arab region. But Facebook users remained limited to younger generations between the age of 18 and 24 years of age (41 percent of the total number of internet users).

Since 2011, the political situation in MENA has been volatile and hostile at different levels. The young digital activists have succeeded in fostering budding dialogues about the role of IT for political mobilization towards regime change and pro-democracy movements (Hofheinz, 2011). Yet this IT euphoria led to misinformation about the reality of digital politics and was commonly overtly simplified with conclusions about the role of social media and the Internet in fostering the struggle for freedom.

In a semantic search based of Facebook posts during the 2011 uprising, keywords related to the protests were examined to measure how much time it took for posts to result in responses like comments. In November, the average was four days, but the following day after Bouazizi burned himself in Tunisia, it only took eight hours, and in January it was only two hours, and when the former president Ben Ali fled the country, it only took three minutes. This increase in frequency was associated with the emergence and use of new terms such as "Streetbook," that served to explain "the transfer of the interaction from social networks to manifestation in the real world, on the street was speeding up (Garbaya, 2011)."

And according to the *Arabic Network for Human Rights Information*, the most popular hash tags between January and March 2011 were #Egypt (1.4 million mentions), #jan25 (1.2 million mentions), and #Libya (990,000 mentions). There were (35,000) active blogs in 2009, which grew to (40,000) by 2010. Since the 2011 uprisings, which is estimated at (600,000) blogs today.

In another study, which mapped how collective memories was shaped and redefined in reference to the online encyclopedia (Wikipedia), results indicated that Wikipedia was first used to advance diverging positions, while using strong emotional narrative to the traumatic happenings. Gradually, the emotional heat was softened and was transformed into a more logical or neutral expression of pain and happenings. The number of direct contributions to the articles decreased, and a substantial part of them became minor edits or indirect work for maintenance purposes (Ferron & Massa, 2011).

Having said so, online activism in North Africa is rapidly evolving and with one of the highest global per capita internet penetration levels, which has grown by (1500 percent) over the last decade, and a dynamic young population is quickly emerging as one of the highest in the world. This marks the region with eagerness to embrace 'newness' (Kaspersky Lab, 2014).

It also means that almost all governments acknowledge the importance of IT and have invested heavily in this sector, through they remain skeptical over its use by opposition movements. Nikita Shvetsov, Chief Technology Officer at Kaspersky Lab, explained that there are many unresolved questions concerning he legal framework governing the processing, storage and transfer of data raise many concerns (Kaspersky Lab, 2014).

Having said so, one could observe that the recent surge in digital activism in MENA was confronted with very stiff resistance by governments with the aim of reducing the impact of digital organizing and independent media. Many regional governments continue to use a variety of technological and offline strategies to go after online media and digital activists.

In Tunisia, Internet service providers logged usernames and passwords to hack into and dismantle online organizing and information sharing among protesters. In early June 2011, Google reported a phishing attack targeted at military and human rights activists to gain access to their Gmail accounts. In Syria, the Syrian Electronic Army carried out attacks to disable and compromise web sites that are critical of the Syrian regime (Faris, et al., 2011).

Many of these attacks involved a mix of private sector and governmental actors, which blur the traditional lines between cyber attacks and government surveillance. In such an environment, maintaining online security is a growing challenge. This might explain the reasons behind the position of the region with higher than the global average with more than

(23 percent) and even outpaced the global average of (17 percent) that makes it vulnerable to outside attacks.

Unfortunately, most digital activists are vulnerable to all types of such attacks due to poor security awareness and education programs with absence of "privacy and cyber legislation" which make the governments' job easier (Bisto, Fourie & Bothma, 2010).

It is thus not surprising to find authorities contacting a content producer or host, by telephone, and ask a particular information be deleted from the internet. In some cases, individual bloggers or webmasters are threatened with various reprisals, if they fail to comply with what the authorities wanted! Over the past years, activists from Egypt, and Tunisia found that their YouTube videos or Facebook accounts removed or disabled after complaints were filed, apparently by regime supporters (Kelly & Cook, 2011).

Conclusion

Over the years, IT contributed greatly to the success of establishing new spaces that could establish and sustain democracy in North Africa. Such unprecedented space, tools and potentials enhanced the quality of data collected, as well as making it possible to identify a roadmap for progress and platform for engagement.

Despite this break through of IT, the long list of challenges remained in place; including poor infrastructure, lack of training, and resistance to change from governments and civil society alike have all prevented ICTs from realizing their true potential.

It seems odd that North Africa has yet failed to convert its heritage into a cultural industry that limited the potential role and impact of its geo-cultural informatics map, when it comes to its contribution and participation are not transformed into a knowledge society.

North Africa like the rest of the content still faces economical, political and cultural deficiencies, in particular with the predominance of primordial societies that buries unlimited opportunities in static traditions, as well as fundamentalist and nationalist demagogies.

It is still difficult to establish a strong relationship of causality between IT and political revolutions. It is thus very timely to figure out the complexities involved such as the network of events, forces, and people in order for social media to be effective in political change, by acknowledging the differences in content and capability among various forms of IT and the level of accessibility, motivations and skills entailed.

North Africa lives in "youth bulge" with new IT users who became more aware about class and gender inequalities as well as the transformative impact that IT can contribute in their lives and futures. Youths are increasingly anchored in the everyday challenges, though they found their innovative and creative mechanisms in expressing their dissent and challenge the traditional establishments of the military or formally entrenched socio-cultural norms, as well as rebel against the prevailing political order in their countries.

The chapter revealed some of the digital practices of thee "Youth Bulge," however there remains push and pull factors that shape the continuous interaction between states, civil society and technology. IT cannot have revolutionary or political ends!

To end, there are two key observations in this context: first, the online world is not entirely cut-off from real life, which includes the kind of power struggle and cultural responsiveness to political happenings. The second observation related to how IT can serve anti-democratic ends, by advancing remote sensing and invasion of privacy, which requires a new kind of thinking, a new domesticated theories and more insight into the context. Otherwise, the academy will remain entrapped in its own predispositions.

REFERENCES

Adeiza, M. (2013). The trouble with social media in Africa. *New African*, (March 2013), pp. 78-79.

Barber, B.(2001). The Uncertainty of Digital Politics: Democracy's Uneasy Relationship with Information Technology. *Harvard International Review*, (May, 6, 2001). Available At: http://hir.harvard.edu/mediathe-uncertainty-of-digital-politics/.

Barkawi,T. (2011). The globalization of revolution. *Al-Jazeera,* Opinion. Available At: http://english.aljazeera.net/indepth/ opinion/2011/03/201132013193456857З.html.

Baym, N. (2010). *Personal Connections in the Digital Age.* Cambridge: Polity.

Benson, R. (2005). The political shaping of the media. *European Journal of Sociology,* XLVI (3), 541–544.

Bisto, C., Fourie, I. & Bothma, T. (2010). Trends in transition from classical censorship to Internet censorship: selected country overviews. *IFLA.* Available At: http://www.ifla.org/files/assets/faife/publications/spotlights/1%20Bitso_Fourie_BothmaTrendsInTransiton.pdf.

Castelle, M. (2011).A Network Theory of Power. *International Journal of Communication,* 5 (1),773–787.

Castells, M. (2004). *The Power of Identity. The Information Age: Economy, Society and Culture,* 2(1), 2nd edn. Malden, MA: Blackwell.

Carty, V. (2010). *Wired and Mobilizing: Social Movements, New Technology, and Electoral Politics*. London: Routledge.

Chadwick, A. (2008). *Web 2.0: New Challenges for the Study of E-Democracy in Era of Informational Exuberance*. ISJLP, 5, 9.

Dahlberg, L. (2007). Rethinking the fragmentation of the cyber public: from consensus to contestation. *New Media & Society*, 9(5), 827–847.

Davenport, C. (2010). Rashomon effect, observation, and data generation. *Media Bias, Perspective, and State Repression.* Cambridge University Press.

Davison, P. (1983). The Third-Person Effect in Communication. *Journal of Public Opinion Quarterly*, 47 (1), 1-15.

Eaton, T. (2013). Internet Activism & the Egyptian Uprisings: transforming online dissent into the offline world. In *Westminster Papers in Communication Culture: The role of social media in the Arab uprisings, past and present*, 9(2),3-24, (April 2013).Available At:

https://www.westminster.ac.uk/__data/assets/pdf_file/0004/220675/WPCC-vol9-issue2.pdf.

ECOSOC Annual Ministerial Review, (2013). Science, Technology and Innovation for Sustainable Development. *2013 ECOSOC Annual Ministerial Review, Regional preparatory meeting for Western Asia.* Available At: http://www.un.org/en/ecosoc/newfunct/pdf13/amr-jordan-background.note.pdf.

Eickelmand, D. & Anderson, J. (2003). *New Media in the Muslim World: The Emerging Public Sphere.* Bloomington, IN: Indiana University Press.

El Guindy, M. N. (2014).*Cybercrime Legislation in the Middle East Cybercrime Legislation in The Middle East.* Available At: http://webcache.googleusercontent.com/search?q=cache:GyhNBhvxuuQJ:www.slideshare.net/askpc/cybercrime-lawmen+&cd=5&hl=en&ct=clnk&gl=za.

El-Nawawy, M. & Khamis, S. (2012). Political Activism 2.0: Comparing the Role of Social Media in Egypt's "Facebook Revolution" and Iran's "Twitter Uprising". *Online Journal of the Virtual Middle East*, 6(1). Available At: http://www.cyberorient.net/ article.do?articleId=7439.

Emminghaus, W. B., Kimmel, P. R. & Stewart, E. C. (1998). Primal Violence: Illuminating Culture's Dark Side." In *The Handbook of Interethnic Coexistence*, ed. Eugene Weiner, (New York: Continuum Publishing.

Faris, R., Roberts, H., Heacock, R., Zuckerman, E. &, Gasser, U. (2011).Online Security in the Middle East and North Africa: A Survey of Perceptions, Knowledge, and Practice. *Cyber Law Harvard*, (August 2011), Available At: http://cyber.law.harvard.edu/sites/cyber.law.harvard.edu/files/OnlineSecurityintheMiddleEastandNorthAfrica_August2011.pdf.

Franklin, M. I. (2004). *Postcolonial Politics, the Internet and Everyday Life: Pacific Traversals Online*. London: Routledge.

Fraser, N. (1997). *Justice Interruptus: Critical Reflections on the 'Postsocialist' Condition.*New York: Routledge.

Ferron, M. & Massa, P. (2011). Wiki Revolutions: Wikipedia as a Lens for Studying the Real-time Formation of Collective Memories of Revolutions. *International Journal of Communication*, 5 (1), 1313–1332.

Garbaya, S. (2011).The other dimension of the virtual space in the revolution of freedom in Tunisia: from Facebook to Streetbook. *The North Africa Journal,* 225(1), (March 14, 2011). Available At: http://www.north-africa.com/.

Ghannam, J. (2011). Social Media in the Arab World: Leading Up to the Uprisings of 2011. *A Report to the Center for International Media Assistance,* (February 3, 2011), Washington, D.C.

Gibbon, P. Durkin, K. (2006). The third person effect: Social distance and perceived media bias. *European Journal of Social Psychology*, 25(5), 597–602.

Habermas, J. (2006). Political communication in media society: does democracy still enjoy an epistemic dimension? The impact of normative theory on empirical research. *Communication Theory*, 16(1), 411– 426.

Hindman, M. (2009). *The Myth of Digital Democracy.* Princeton, NJ and Oxford: Princeton University Press.

Herrera, L. (2015). Citizenship under Surveillance: Dealing with the Digital Age. *Int. J. Middle East Stud.,* 47 (1), 354-356.

Hofheinz, A. (2011). "Nextopia? Beyond Revolution 2.0." *International Journal of Communication* 5 (1), 1417-1436.

Howard, P. N. (2011). *The Digital Origins of Dictatorship and Democracy: Information Technology and Political Islam.* Oxford: Oxford University Press.

Hussain, M. M. (2014). Digital Infrastructure Politics and Internet Freedom Stakeholders after the Arab Spring. *Journal of International Affairs,* 68 (2014), 37–56.

Idlebi, N. (2015). Expert Meeting on Cyber laws and Regulations for Enhancing E-commerce: Including Case Studies and Lessons Learned. *Development of Cyber Legislation in the Arab Region*, (25-27 March

2015). Available At: http://unctad.org/meetings/en/Presentation/CII_EM5_P_NIdlebi_en.pdf.

Kahn, R. and Kellner, D., (2004). New media and internet activism: from the 'Battle of Seattle' to blogging. *New Media & Society*, 6 (1), 87–95.

Christopher F. Karpowitz, C. F., Raphael, C. & Hammond, A.S. (2009). Deliberative Democracy and Inequality: Two Cheers for Enclave Deliberation among the Disempowered.*Politics & Society,* (December 2009), 37(1), 576-615.

Kaspersky Lab, (2014). Kaspersky Lab Technologies Extend Powerful Real-time Data Protection to Hitachi NAS Platform Systems. *Kaspersky Lab*, (Nov. 25, 2014). Available At:http://www.kaspersky.com/about/news/business/2014/Kaspersky-Lab-Joins-Hitachi-Technology-Alliance-Program.

Kelly, S. & Cook, S. (2011). New Technologies, Innovative Repression: Growing Threats to Internet Freedom. *Freedom House Report, Freedom on the Net.* Available At: https://freedomhouse.org/report/freedom-net-2011/new-technologies-innovative-repression-growing-threats-internet-freedom.

Kern, K. (2008). *Heterotopia of the theme park street. In: Dehaene M and de Cauter L (eds) Heterotopia and the City.* London: Routledge.

Kreutzer, T., (2009). Generation mobile: online and digital media usage on mobile phones among low- income urban youth in South Africa [online]. *Complete survey results.* Available At: http://tinokreutzer.org/mobile/MobileOnlineMedia-SurveyResults-2009.pdf.

Laclau, E. & Mouffe, C. (2001). *Hegemony and Socialist Strategy: Towards a Radical Democratic Politics.* (2nd edn). London:Verso.

Lievrouw, L.A. (2011). *Alternative and Activist New Media.* Cambridge: Polity.

Mabweazara, H. M. (2010). "New" technologies and journalism practice in Africa: Towards a critical sociological approach. In N. Hyde-Clarke (Ed.), *The citizen in communication,* (pp. 11–30). Claremont, CA: Juta.

MacKinnon, R. (2012). *Consent of the Networked: The Worldwide Struggle for Internet Freedom.* New York: Basic Books.

McCaughey, M. & Ayers, M.D., (eds.), (2003). *Cyber activism: online activism in theory and practice*. New York: Routledge.

Mesbahi, M. (2007). The Third World and the paradox of the digital revolution. *International Review of Information Ethics*, 7(4), (Sept. 2007). Available At: http://www.i-r-i-e.net/inhalt/007/04-mesbahi.pdf.

Morozov, E. (2012). *The Net Delusion: The Dark Side of Internet Freedom Paperback use pre formatted date that complies with legal requirement from media matrix*. New York: Public Affairs.

Movius, L. (2010). 'Cultural Globalisation and Challenges to Traditional Communication Theories.' Platform: *Journal of Media and Communication*, 2(1), 6-18.

Nichols, J., & McChesney, R. W. (2009). The death and life of great American newspapers. *The Nation*, April 6.

Obadare, E., (2004). 'The Great GSM Boycott: Civil Society, Big Business and the State in Nigeria.' *Dark Roast Occasional Paper Series*, Isandla Institute.

Osundare, N. (2002). *Thread in the Loom: Essays on African Literature & Culture*. Trenton, N.J: Africa World Press.

Owen, T. (2015).Disruptive Power: The Crisis of the State in the Digital Age. In *Oxford Studies in Digital Politics*, (28 May 2015). Oxford Press.

Postill, J. (2012). Digital politics and political engagement. In H. Horst and D. Miller (eds) *Digital Anthropology*. Oxford: Berg.

Rao, M., (2011). Mobile Africa report: regional hubs of excellence and innovation. *Mobile Monday,* [online]. Available At: http://www.mobilemonday.net/reportsMobileAfrica_2011.pdf.

Roberts, B., (2009). Beyond the 'networked public sphere': Politics, participation and technics in Web 2.0. *Fibreculture Journal,* 14(1). Available At: http://journal.fibreculture.org/issue14/issue14_abstracts.html.

Salmon, L. Fernandez, & Post, L. A. (2010). Mobilizing public will across borders: roles and functions of communication processes and technologies. *Journal of Borderlands Studies*, 25(1), 159-169.

Sarrazin, T. (2011). Texting, Tweeting, Mobile Internet: New Platforms for Democratic Debate in Africa. *Fes Media Africa Series,, Friedrich-Ebert-Stiftung.* Available At: http://archive.kubatana.net/docs/media/fesmedia_new_media_platforms_110727.pdf.

Schraeder, P. J. (2003). *African Politics and Society: A Mosaic in Transformation*. Belmont, CA: Wadsworth Publishing, 2nd ed.

Shenker, J. et al. (2011) Young Arabs who can't wait to throw off shackles of tradition. *The Guardian,* (14 February, 2011). Available At: http://www.guardian.co.uk/ world/2011/feb/14/young-arabs-throw-off-shackles-tradition.

Soyinka-Airewele, P. (2010). Canons of Conformity, Cannons of Audacity: Insurgency and Voice in the Emergent African Cinema. *Paper Presented at the African Speaks 2010 Visiting Lecturer Series At the University of South Africa*, Pretoria Monday, (March 29, 2010). Available At: http://www.unisa.ac.za/contents/colleges/col_humanities_social_sciences_education/docs/canons-of-conformity.pdf.

The Economist, (2012). Online activists in the Middle East: Seeds of the future. *The Economist,* (May, 22, 2012). Available At: http://www.economist.com/blogs/newsbook/2012/05/online-activists-middle-east.

The Global Evolution of Digital Commerce and MENA eCommerce Report, (2013). *eCommerce report commissioned by VISA for the Middle East and North Africa (MENA)*, The International Omni Retailing Members Association (IORMA), (April, 2013). Available At: http://www.visamiddleeast.com/me/common/include/uploads/ecommerce_apr2013.pdf.

Tung, E.(2011). Social Networks: The Weapons of our Modern Era. *The Talon,* (February 28, 2011). Available At: http://my.hsj.org/Schools/Newspaper/tabid/100/view/frontpage/schoolid/3302/articleid/418099/newspaperid/3415/Social_Networks_The_Weapons_of_our_Modern_Era.aspx.

United Nations Office on Drugs and Crime (UNODC), (2013). *Comprehensive Study on Cybercrime.* Vienna: United Nations.

US Department of State, (2012). Freedom Online Conference Launches Digital Defenders Partnership. *Media note, Washington*, D.C., (13

September 2012). Available at: http://www.state.gov/r/pa/prs/ps/2012/09/197723.htm.

Vehg, S. (2003). Classifying forms of online activism: the case of cyberprotests against the World Bank. In: McCaughey M and Ayers MD (eds) *Cyberactivism: Online Activism in Theory and Practice.* London: Routledge, pp. 71–96.

Wiktorowicz, Q. (2004). *Islamic Activism: A Social Movement Theory Approach.* Bloomington, IN: Indiana University Press.

Women Learning Partnership (WLP), (2013). Identity, Gender, and Activism in the Middle East-North Africa: An Online Discussion with Activists & Artists from the Region. *WLP,* (December, 10, 2013). Available At: http://www.learningpartnership.org/identity-gender-activism-mena-event.

Zald, M. N. & McCarthy J. (1987). *Social Movements in an Organizational Society.* New Brunswick, NJ: Transaction.

INDEX

D

E

R

S

T

U

V

W

X

Y

Z